The Merlin

The Merlin

The Engine that Won the Second World War

Gordon A. A. Wilson

AMBERLEY

To F. H. (Henry) Royce who conceived the engine, to those who developed the engine, to those who maintained the engine and to those who flew the Merlin to victory.

Merlin, the bird, illustration title page by H. R. (Rick) Hovey.

First published 2018

Amberley Publishing
The Hill, Stroud
Gloucestershire, GL5 4EP

www.amberley-books.com

British Library Cataloguing in Publication Data.
A catalogue record for this book is available from the British Library.

ISBN 978 1 4456 5681 6 (hardback)
ISBN 978 1 4456 5682 3 (ebook)

Typesetting and Origination by Amberley Publishing.
Printed in the UK.

Contents

Acknowledgements

My wife Emily for her love, editing and continued support of my writing career. My family for their love and encouragement. Peter Collins, Chief Executive Officer, Rolls-Royce Heritage Trust, for being so helpful and providing necessary images for the Merlin story. Jose Flores, Steve Hinton, Greg Morrison, Andrew Panton, Martin Rouse, the interviewees who took the time to speak with me, thank you. Gary Vincent for research, access to his aviation library and his world-wide photography contacts for superb images. You saved the day! Barrie Gabie for his technical explanations, fact checking and grammatical editing. Barrie Laycock for editing and research.

Colm Egan for website research and image adjustments. Geoffrey Pickard for use of his aviation library. Wayne Ralph (Author & Military Historian) for use of his aviation library. H. R. (Rick) Hovey for the merlin bird frontispiece. Anne Tilbury, Information Technology support.

Richard Folgar, Archivist Air Canada, image permission. David Birrell, Archivist & Librarian for research access, Bomber Command Museum of Canada. Richard de Boer, Calgary Mosquito Society. Douglas Bowman, Archer Photo Works. Surrey Public Library, Inter-Library Loan department.

Images courtesy of: Air Canada Archives, John Allen, Bomber Command Museum of Flight, Douglas Bowman, Calgary Mosquito Society, Canadian Museum of Flight, Canadian Warplane Heritage Museum, David Birrell, Richard de Boer, Keith Burton, K.C. (Colm) Egan, Atushi 'Fred' Fujimori, Steve Hinton, Huyghe-Decuypere Group, John Kimberley, John Mounce, Malcolm Nason, Eddie O'Brien, Geoffrey Pickard, Bill Powderly, Wayne Ralph, Martin Rouse, Rolls-Royce Heritage Trust, Jennifer Stevens, Richard Vandervord, Jerry Vernon.

Many thanks to the staff of Amberley Publishing for their work on *The Merlin* especially, alphabetically, Shaun Barrington, Alex Bennett, Jonathan Jackson, Cathy Stagg and former employee Jonathan Reeve. Without you all, the Merlin would have remained in 'literary silence'.

My sincere apologies to those I may have missed. Rest assured your contributions were much appreciated and contributed to the Merlin story and any such omissions will be rectified on reprint.

Valuable use for research and ideas was made of the series of publications by the Rolls-Royce Heritage Trust, the Sir Henry Royce Foundation and Peter Pugh's *The Magic of a Name: The Rolls-Royce Story* Parts One, Two and Three, and Ian Craighead's *Rolls-Royce Merlin*, all listed in the bibliography.

Foreword

The Rolls-Royce Merlin is one of the most significant aero-engines to have been produced, powering some of the most iconic aircraft of the 20th century. Tracing its origins to the Schneider Trophy races, the Merlin engine soon became a legend in the skies over the European and Pacific Theatres during the Second World War; even today this engineering marvel is still pushing the boundaries of piston-powered aviation.

My first taste of aviation came at the age of two weeks when I attended the Reno National Championship Air Races; there, my father piloted the purpose-built air racer known as *Tsunami*, powered by none other than the Rolls-Royce Merlin engine. As a child much of my youth was spent growing up around Chino Airport and the Planes of Fame Air Museum, whose collection houses a variety of aircraft, including many Schneider Cup aircraft replicas in addition to a few North American Aviation P-51 Mustangs. At the age of sixteen I had another epiphany at the Reno Air Races when my eyes fell on a beautiful red-and-white P-51 Mustang named *Strega*. Her aesthetically pleasing aerodynamics were mesmerising, and the sound which screamed out of the V-12 engine enchanted me like the call of a siren; I became enamoured of the Merlin-powered racer and my life was forever altered.

When I returned home I purchased a soundtrack of the Reno Air Races and became addicted to the howl of the Merlin engine, whose output now exceeded 3,400 horsepower, a far cry from the original 1,500 horsepower Mustangs and Spitfires nearly four score years ago. *Strega* was kept in Bakersfield, California, where I met the aircraft's crew chief and began volunteering my time on weekends to help prepare the Mustang to compete in Reno. Much to my delight, he offered me a position on the crew to help maintain and assist with all facets of the racing campaign, which quickly became a year-round endeavour.

As a couple of years passed, I found myself qualified in the P-51 Mustang at the age of nineteen; my father's P-51's origins stemmed from the *Red Baron* racing aircraft with which he set the 3km World Speed Record for Piston-Powered Aircraft, and this was where I cut my teeth. At the same time, Mike Nixon and Jose Flores of Vintage V12s invited me to work as an intern in their engine shop, which is world renowned for the overhaul and maintenance of Rolls-Royce engines, among other types. I was fortunate to have had that opportunity and I've never learned so much in such a short amount of time. They gave me considerable insight to the Merlin engine, and it wasn't long before I was disassembling engines in their entirety; I learned the procedures for proper maintenance and overhauls, as well as how to troubleshoot, diagnose and maintain a reliable engine.

These skills transferred nicely to air racing, where much of the maintenance during the race week pertains to ensuring the Merlin engine is healthy, changing magnetos, torqueing the heads and banks, adjusting valves and retiming the engine. Around that time, the stars aligned for me and I was given the opportunity to take over the piloting duties of *Strega* during the air races. Thanks to the efforts of countless individuals, and the Rolls-Royce engine in front of me, I was very successful and won the National Championship a handful of times.

After a stint with *Strega*, I was recruited to pilot and campaign another P-51 called *Voodoo*. I was fortunate to have also had much success with this aircraft, and achieved a lifelong dream of establishing a new speed for the 3km World Speed Record of 531.65 mph. Of all the aircraft that have set the record and raised the bar since 1939, none were powered by an engine of less than 2,000 cubic inches; the Rolls-Royce Merlin that propelled *Voodoo* and me to a new record was, of course, *1,650* cubic inches. I believe this speaks volumes for what the engine is capable of achieving; a testament not only to the men and women who maintain these marvels today, but a substantial tribute to the engineers who designed such a masterpiece.

The Rolls-Royce Merlin engine's history has certainly come full circle; from its birth out of air racing, to gaining air superiority for the Allies in the Second World War, and finally the modern era when it again asserts dominance over the Nevada desert, holding its proper place as the fastest piston engine in the world. Gordon's book is the ultimate testament to the Rolls-Royce Merlin and is truly an everlasting legacy, exemplifying what mankind can accomplish with a common goal.

I hope that each of you may have an opportunity to enjoy the music of the Merlin, a symphony of mechanical components which sing a song so distinct that whoever hears it will become enchanted by it, just as I still am today.

Happy reading.

Steve Hinton Jr

(Steve Hinton Jr was the unlimited-class champion of the Reno Air Races in 2013, 2014 and 2016 flying the modified NA P-51 Mustang *Voodoo*.)

Introduction

The pilot hauled the Lancaster in a tight turn over the dark hills and felt the surge of power as he advanced the four throttles and dove the aircraft to 60 feet above the darkened lake surface. It was night. It was dangerous to fly so low over the dark water but it was war. It was 1943 and he was deep in Germany leading a bombing mission to destroy several dams in the Rhine Valley, the very heartland of the enemy's industrial region. The pilot was Wing Commander Guy Gibson, who would later be awarded the Victoria Cross (VC) for this raid, leading the 617 'Dam Buster' Squadron Royal Air Force (RAF) on this special operation. Top-secret in its planning, the mission was to destroy the Mohne, Sorpe and Eder dams with a specially designed bomb that required release at a precise altitude and airspeed. This famous event is immortalised in the 1955 film *The Dam Busters* starring Sir Michael Redgrave CBE and Richard Todd OBE, and in many books.

The mission was termed a success, but it was not without loss of aircraft and lives among the seven-member crews. It would take a special crew and a special aircraft to carry out this challenging and dangerous mission. The special aircraft chosen for this unique mission was the rugged Avro Lancaster heavy bomber. The fuselage of the aircraft had been modified to carry the special rotating bouncing bomb designed by Barnes Wallis. The Dam Buster Lancasters carried their bomb load over a lengthy, circuitous route, defended themselves against enemy fighters, manoeuvred dangerously low at night to deliver the bombs and returned to their home base in England. The fuselage, wings and empennage are the 'body' of the aircraft, but the engines are the 'heart'. The Lancaster has plenty of heart because of the four powerful Rolls-Royce Merlin engines. The reliable Merlin got the surviving crews safely home.

The Merlin was one of the tools that brought the Lancaster back to its RAF station, using the word 'tool' in its broadest terms. Since the beginning of time man has used tools or simple machines to make his work easier and more efficient. The tool had evolved from a hammer to, among many other devices, a powerful aircraft engine in the 1930s that converted one form of energy into a force to achieve work. The Merlin took combustion energy and turned it into mechanical energy using a propeller. The propeller then converted the mechanical energy to a form of energy known as motion. The Lancaster now had something, four famous somethings, the Merlin engines, to propel it through the air, and the term airspeed to describe how fast the aircraft was passing through the air.

The flight of birds had been observed for aeons and man wished to join them in their aerial environment for ease of travel over rough terrain, to see prey at long distances and no doubt to enjoy the freedom of flight. Leonardo da Vinci drew some human-powered flying machines with flapping wings back in the fifteenth century. The problem was that the human did not have enough power to make the machine work. Interestingly, da Vinci also drew some gliding fixed-wing flying machines, so his thought process was going in the right direction.

The engine transformed the very restricted flying machine, the glider, into a bird-like machine capable of altitude and – more importantly – distances. Weight, power, size and reliability were the challenges to the aircraft engine designer. The Wright Brothers broke the first and most important flight barrier – a heavier-than-air machine capable of flight. That was more than 110 years ago, and only thirty years before the Merlin entered the aviation scene.

The Wright engine that permitted man to fly was not very sophisticated compared to other piston engines available at the time to the automobile industry. It was a four-cylinder inline horizontal engine with a 4-inch bore (diameter of cylinder) and a 4-inch stroke (distance between the bottom and top of piston movement). It was designed and built by Charlie Taylor, a mechanic hired by the Wrights to work in their bicycle shop. The cast-iron cylinders fitted into a cast-aluminium crankcase, which was unique at that time. The crankcase extended outward, creating a water jacket to cool the cylinders. The water was not recirculated through a radiator but simply replaced the water in a tray that evaporated due to the heat of the engine.

The 12-horsepower engine was sufficient to propel the Wright Flyer I fast enough to create lift with enough horsepower to overcome aerodynamic drag once airborne. It used contra-rotating propellers to achieve the required energy. The automobile-inspired engine could now defy gravity and fly like the birds. The world would never be the same. The United States of America (USA) was not the only country interested in powered flight, just the first to achieve it. Shortly after 1903, other countries and inventors conquered the barrier of heavier-than-air vehicles flying under their own power and the aviation industry came of age. Frenchman Louis Bleriot, for example, became inspired by seeing Clement Ader's Avion III at the 1900 *Exposition Universelle* in Paris. Bleriot became the first person, in 1909, to fly across the English Channel in his Bleriot XI monoplane powered by a 25-horsepower Anzani three-cylinder fan configuration (semi-radial) engine.

The aviation engine design was changing rapidly as engineers and inventors sought to design the perfect power plant. The engine fell into two broad categories: the air-cooled or liquid-cooled engine. The manufacturer then had to decide on the configuration based on the airframe requirements. The inline engine, upright, inverted or horizontal, had many alternatives such as the radial, semi-radial or fan, rotary, V-shaped and the opposed cylinder engine. Each manufacturer had their preferences; keeping in mind the experimental aspect of aero engines 100 years ago. The engine was then matched to a 'suitable' propeller, itself very much a work in progress.

There were many, many, engine manufacturers in the race to conquer the 'power-to-weight ratio' challenge at the beginning of the last century. Some of them were The Curtiss Aeroplane and Motor Corporation in the USA; the Antoinette engine was designed and built in France by Léon Levavasseur; the Anzani was an engine manufacturer founded by Italian Alessandro Anzani, which produced proprietary engines for aircraft in factories in Britain, France and Italy; the Gnome 50-horsepower rotary engine revolutionised aviation and was used extensively in airplanes during the first years of World War I. Le Rhône and Clerget rotary engines, both built in France, were used in many Allied fighter planes. The Spanish Hispano-Suiza liquid-cooled V-12 was developed in 1915 by a Swiss engineer, Marc Birkigt. Among these engine manufacturers vying for market share and contributing to the war effort was the British Rolls-Royce Company with its Eagle engine, which marked the beginning of a line of aviation engines that produced the famous Merlin of the Second World War.

The Rolls-Royce Company was founded by two visionaries who saw that by combining their talents in a business venture, the outcome held the potential for financial success. Charles Rolls and Henry Royce were the two optimists from very different backgrounds. Maybe that is what made the venture a success although one of them sadly, Charles Rolls, would lose his life prematurely in a flying accident.

The Hon. Charles Rolls was the third son of Lord Llangattock, and spent his youth on the family estate in Monmouthshire. Rolls completed his university education and subsequently became interested in motor cars, balloons and powered aircraft. He formed a company that sold cars of quality, mainly French, to his aristocratic friends. Rolls deplored the fact that there was no British car that would meet his exacting standards.

Royce, on the other hand, came from humble beginnings and one of his motivations was to provide for his family and improve their lifestyle. He became interested in electricity and later formed a company with a partner to manufacture quality electrical components. Royce had purchased a French car and set about improving it in favour of designing his own 'better mousetrap'. He insisted on engineering perfection and the first Royce car was exactly what Rolls was looking for, a British-built quality car. Rolls signed an agreement to take the entire output of the Royce Company and the rest, as they say, is history. The name of Rolls-Royce was synonymous with quality, style and performance.

The Rolls-Royce factory at Derby ceased producing cars with the advent of the First World War. It did, however, convert some remaining chassis to armoured vehicles for the war effort. The government required Rolls-Royce to produce aero engines of French and government designs. Royce objected to his company building what he thought were inferior products and designed and produced the Eagle aero engine. Rolls-Royce adopted the nomenclature of birds of prey for its piston aero engines. The Eagle developed 225 horsepower and was a V-12 liquid-cooled design. It powered both fighters – the Farman Experimental 2b – and bombers, the Handley Page 0/400. When hostilities ceased in 1918, Rolls-Royce had built 60 per cent of all British engines.

The company now turned its attention back to producing superior motor cars. The staff doing design, engineering and production, as well as the skilled tradespeople, were all still employed at Rolls-Royce and the company could easily switch focus. However, the wartime experience of building aero engines was not forgotten; engine development continued on a less urgent basis. In the mid-1920s a new sound came from the engine test bed. It was the high-revving sound of the Falcon X, a compact all-aluminium V-12 liquid cooled engine. This was the start of a very long successful line of aero engines, no doubt influenced in the 1930s by the impending prospect of war. The Falcon X became the Kestrel and so continued the engine design challenge of the power-to-weight ratio. Aluminium was a natural choice.

The Schneider Trophy Race in 1931 hastened the design of the R engine from the H engine, the Buzzard. On the day of the race the R engine achieved 1900 horsepower. Supermarine and Rolls-Royce combined to win the Schneider Trophy in 1931 for the third time in a row and Great Britain got to keep the trophy in perpetuity. The R engine was also used by Sir Malcolm Campbell to achieve land and sea speed records for Great Britain. In 1933, the PV (Private Venture) -12 engine first ran on a test bed and, by 1936, was in full production renamed as the Rolls-Royce Merlin. The most Merlins produced were for the Avro Lancaster bomber but, generally, the engine is associated with the British wartime icon, the Supermarine Spitfire.

Chronologically, the Merlin first flew in the Fairey Battle, followed by the Hawker Hurricane, and Supermarine Spitfire. It then flew in more than forty different aircraft types, including flying test beds and post-war military and civilian aircraft. Bomber aircraft included the Armstrong Whitworth Whitley, Avro Lancaster, Avro Manchester III, Handley Page Halifax, Vickers Wellington Mk II & Mk VI and the multi-use de Havilland Mosquito. Fighter aircraft included the Bristol Beaufighter II, Hawker Hurricane, North American Mustang Mk X and the Supermarine Spitfire.

After the Second World War, the Rolls-Royce Merlin powered the Avro York and Canadian North Star military transports. It also powered the Avro Lancastrian and Avro Tudor commercial airliners. The Merlin was subsequently developed as a tank engine and, with the supercharger replaced by a carburettor, was renamed the Meteor.

What sort of organisation could produce the Merlin engine to power these famous aircraft? Who were the men who made the whole manufacturing process possible? How did Rolls-Royce take the designers' ideas and achieve such reliability from a piston engine with so many moving parts? Why was Rolls-Royce so successful in producing the greatest Second World War piston engine ever, the Merlin, to be followed by the greatest turbine engine of the early jet years, the Avon? Rolls-Royce became a globally recognised name from the Arctic to the Antarctic and everywhere in between.

PART 1

THE MERLIN:
BIRTH OF A LEGEND
1904–34
Genesis

1

The Rolls-Royce Company 1904

The Rolls-Royce name conjures up many graphic images. The parade of British royalty being driven down The Mall in London; Maharajahs being driven on the dusty roads of India; chauffeurs dropping off financiers in Wall Street, New York; sheiks being driven on the desert roads of the Middle East; lords and ladies negotiating the twisting country roads back to their English estates; movie stars battling the Los Angles ten-lane traffic on their way to Hollywood; Hong Kong business people parking at the Happy Valley Racecourse for lunch – and so, throughout the world, discerning well-heeled people are enjoying the luxury of the Rolls-Royce car. Many car manufacturers have come and gone in the past 110 years, but Rolls-Royce lives on across the globe as a legend among automobiles. Rolls-Royce would make the ideal car, a beautifully made precision machine, financial feasible.

One only also has to look out the viewing windows of any airport to see the Rolls-Royce logo emblazoned on the engine cowlings of various aircraft manufacturers. The Rolls-Royce Avon engine was 'the Rolls-Royce' of engines in its time. It is remarkable how the name has remained synonymous with quality throughout all these years and is often used as an adjective to describe engines in the transition from the large piston engines to the turbojet engine. The Avon – Rolls-Royce named their jet engines after rivers – was the one of the most successful Rolls-Royce engines during the immediate post-war period and indeed the twentieth century. The Avon was used extensively by the military and also used by the civilian aviation industry. It was two winners in a row for Rolls-Royce, the Merlin, and its jet-age replacement, the Avon.

The company was now one of the giants of the aviation aero-engine industry, along with General Electric and Pratt & Whitney, both in the USA. Competition was fierce as the post-war economic boom drove the aero-engine industry out of the piston era into the jet age. Airlines were expanding rapidly as faster and more efficient aircraft became available. The Cold War ensured the growth of the Royal Air Force, both in the intercept and strike roles. Why was Rolls-Royce nationalised by the British government in 1971 amid all this prosperity? The business model would suggest that this would be a time of sustained growth. We will trace the history of Rolls-Royce from 1906 to try to understand the temporary demise and subsequent recovery of such a widely known name as Rolls-Royce.

Rolls-Royce made its business journey through life using machines and human intelligence. It could be argued that intelligence alone started mankind's journey through the years by taking simple machines and developing them into complicated machines. The simple machine takes a force and changes its direction or its magnitude, in other words, work. The machine was defined by the Renaissance, a period in European history from the fourteenth until the seventeenth century, which marked the beginning of the Early Modern Age. The Renaissance scientists defined the six basic machines as the lever, wheel and axle, pulley, inclined plane, and the wedge and screw. It is interesting to note that the mechanical advantage of the inclined plane was added later to the basic group in the year 1586. Galileo was the first to comprehend that simple machines do not create energy but only transform it to a different form.

We talk about the wheel being 'the greatest invention ever' but on its own it would not amount to that great an influence on the mechanical progression of the world. What really set it apart as 'the greatest invention ever' was when it was successfully matched to the lowly axle; now that was a formidable combination, which would influence many inventions including, for example those in transportation, from the horse and cart to the International Space Station and everything in between. Professor David Anthony of Hartwick College confirms this in his book *The Horse, the Wheel and Language*, when he writes that 'the stroke of brilliance was the wheel-and-axle concept.'

It was an easy transition from Shanks's pony to riding in a wheeled cart pulled by the energy of an animal. At the end of the eighteenth century, 'ordinary folk' generally did not travel far, only adventurers, explorers and commercial entrepreneurs did so. However, it was not long before steam engines were invented to replace 'horsepower' and this was followed by the internal combustion engine. These engines, in combination with the wheel and axle, transformed the world of transportation. Steam ships, buses and railway engines dominated the mass transportation industry on sea and land in the early nineteenth century. The first 'car' was probably a steam-driven tricycle in 1769, which would be progressively developed until the the first modern car was designed by German inventor Karl Benz in 1886. It would take the power of the internal combustion engine on 17 December 1903 in Kitty Hawk, North Carolina, USA, to get a heavier-than-air vehicle, using wheels and axles, to prove the possibility of controlled flight over ever-increasing distances.

So, within the space of seventeen years, from 1886 (Benz) to 1903 (the Wright Brothers), the foundation was laid for the two industries for which Rolls-Royce would become famous, the automobile and aviation businesses. Although there were many internal combustion engines before the twentieth century, it was not until commercial drilling and production of petroleum in the mid-nineteenth century that the petrol or gasoline engine came into its own. It was during this time of rapid invention that Charles Rolls and Henry Royce, from very different backgrounds, met in 1904. Rolls was twenty-seven years old and Royce was forty-one; under the existing rules or conventions of society the two men should never have met. The meeting was a twist of fate that altered their careers and lives and, in this case, the history of luxury cars and aero-engines. Again, as they say, 'the rest is history', and that history we will learn about in the coming chapters. The object of this chapter is to provide a more detailed background to the Rolls-Royce Company and personnel before the end of the Second World War. A general company engine history from post-World War II to the present will be found in Appendix I. The development and production of the Merlin engine will be discussed at length in Chapters 3 to 10.

Frederick Henry ROYCE

The Royce family – father James, a miller by trade, mother Mary, and their four children – moved to a rented mill at Alwalton, Huntingdonshire, in 1858. Frederick Henry Royce was born at the mill on 27 March 1863. The mill had water as well as steam so had the capability of supporting the family. Unfortunately, James suffered from ill health, which caused him to mortgage the lease on the Alwalton mill to the London Flour Co. during the year of Royce's birth. James later went to London, taking his two sons with him, leaving his wife and daughters in Alwalton. After five years, in 1872, he passed away, leaving his son Frederick Henry, known as Henry, then nine years old, on his own in London. At this tender age,

Henry was fending for himself by delivering newspapers for W. H. Smith. His mother had worked nearby in London as his father's health had worsened but was unable to care for him at her home. Royce suffered at that time from cold and hunger, not a very good start in life. Spells of work were interspersed with some education. Royce worked as a Post Office employee in the telegram delivery section in the Mayfair area, where Rolls was born. Years later, this very area would see luxurious motor cars driving by bearing his name. There has been some conjecture that Rolls and Royce had crossed paths before their first recorded meeting in 1904. Royce possibly delivered telegrams of congratulations on Charles' birth to the Rolls' home in 1877. Royce would have been fourteen at the time.

At this time an aunt of Royce's agreed to sponsor him for an apprenticeship with the Great Northern Railway in Peterborough. Royce was reputed to have a fanatical desire to learn more about engineering and to make up for the previous lost years. Unfortunately, his aunt was not able to continue sponsoring him and aged seventeen he scrambled to find work. He was lucky with his limited background to be taken on as a toolmaker with a company in Leeds. While there he developed an interest in electricity, which led him to his next job. Royce joined the Electric Light & Power Co. in London. While there he studied under Professor William Ayrton of the City and Guilds of London Technical Institute and the Central Technical Institute. Ayrton was associated with the faculties of Applied Physics and Electrical Engineering. It is interesting to note that Ayrton worked on the first electric tricycle, so as a backdrop to all his theory, this showed he had a practical interest in replacing the horse-drawn vehicle. Royce also studied under Hiram Maxim, of Maxim machine gun fame who held many other patents, where he studied electric lamps. Royce had responsibility for theatre and street lighting projects in Liverpool, which were completed in early 1884. By May the company had gone out of business.

Royce had saved a little money and a friend, Ernest A. Claremont, one of a series of 'fortunate contacts' during Royce's life, had saved a little more. Claremont also had electrical training and a decision was made in 1884 to form an electrical company in Manchester. Claremont, interestingly, was Royce's only partner; Rolls and later Claude Johnson were co-directors of Rolls-Royce. The driving force of the company venture must have been Royce as the company was called F. H. Royce & Co., electrical engineers, despite Royce having less financial investment in the new company. They rented No. 1a, Cooke Street, Manchester, and this was a defining moment in Royce's continued recovery from the poverty of his childhood. The Honourable Charles Stewart Rolls was seven years old as twenty-one-year-old Frederick Henry Royce embarked on his new life as an independent businessman.

In 1884 electricity was in its infancy and illumination of many public areas was still by gas lamps. Names such as Edison and Swan, among others, Maxim included, were associated with the carbon filament electric light bulb. Patent rights fights were common and provided job security for the lawyers. Joseph Swan was a British physicist and chemist and is famous for inventing the first incandescent light bulb. His home in Gateshead, Tyne and Wear, was the first in the world to be lit by lightbulbs. Other firsts included public lectures and theatre illumination by lightbulbs. In 1881, just three years before Royce formed his company, Swan was awarded the French Legion of Honour and witnessed Paris being lit by his invention, the electric light, during an international exhibition. The light bulb and its technology had been established but without a source of stable power, it was useless. There were few sources of public electrical power available and if an establishment, private or commercial, wanted electric light it would have to have its own generator, which was a costly investment and made electric light a luxury at that time in the late 1880s.

In 1866, Gorges Leclanché invented a wet cell battery that consisted of a zinc anode and a manganese dioxide cathode wrapped in a porous material, dipped in a jar of ammonium chloride solution. This 1.4-volt cell quickly achieved success in telegraphy, signalling and electric bell work. Royce used this battery to provide electric bell sets. The company then progressed to the manufacture of small items such as fuses, switches, bulb holders and instruments. This parts contract business naturally developed into complete installations and further to the manufacture of larger electrical items such as direct current dynamos; he conceived 'sparkless' commutation, and electrical motors that were used in winches and cranes. This development and pioneering work did not come without personal cost, especially for Royce.

Royce was the technical partner and was totally engrossed in his work to the extent that he neglected his own well-being. Lack of sleep and nourishment were common in his working practice. Claremont was the sales and office manager and was a direct contrast to Royce; Claremont was very keen on physical fitness. Today Royce would be referred to as a workaholic. His whole life was devoted to observing, considering and improving what he saw.

Nevertheless, the partnership functioned well both in business and in their social life. In 1889, Claremont, and 1893, Royce, each married a daughter of Alfred Punt, so the two sisters became wives and sisters-in-law to the two business partners. This was unusual and perhaps there was a financial investment in the background, arranged marriages perhaps? Another 'fortunate contact' at the same time was hiring John De Looze as cashier and accountant. De Looze worked with Rolls-Royce until his retirement in 1943; he was both a friend to Royce and a company executive. Ten years after the company was formed, it became a limited company in 1894. In 1897 the Motor Car Club became the Automobile Club of Great Britain with Claude Johnson as the first Secretary. The old boy's network was nicely set up for the future Rolls-Royce, with both Johnson and Claremont moving in the same club circles.

The company was known as F. H. Royce & Co. Ltd. from 1894 to 1899 and became Royce Ltd in 1899. It had outgrown Cooke Street and had built additional premises at Trafford Park in 1901. One of its neighbours was the W. T. Glover electric cable company. One of its products was a lead-sheathed copper core cable for underground use. It was used extensively in the mining business. Henry Edmunds was a partner in the company and managed it for a very successful nine years. He would in fact become the catalyst in another partnership of great importance. Royce had now built a home in Knutsford, which had its own electrical installation and must surely have been the envy of his neighbours and further enhanced his reputation as an engineer.

Royce had purchased a De Dion quadricycle and so began his life-long association with evolving transportation, which would eventually become the car. No doubt, as an engineer, he quickly saw improvements that could be made to the De Dion design.

In 1902 Royce collapsed from exhaustion, his body could no longer take the unlimited hours and lack of care that he imposed on himself to establish the new business location. Cheap foreign-made parts from the USA and elsewhere were taking their toll on the home business front. His wife had relatives in South Africa and a subsequent ten-week trip there somewhat restored his health and energy levels. It was a slightly strange choice of destination as the Boer War had just ended in May 1902.

Modes of transport were changing fast and the next logical choice for a partner in a progressive engineering company would be to change the quadricycle for a car. Royce chose the 10hp twin cylinder Decauville. Royce possibly chose this particular make of car because the Decauville manufacturing company, founded by Paul Decauville, specialised in industrial railways before getting in to cars. Decauville's innovation were ready-made sections of light narrow-gauge track that were easily transported and assembled. The 10hp car had the company reputation of reliability behind it.

Royce, however, was not satisfied with the quality and reliability of the car; it did not live up to its associative railway reputation. Rather than designing electric motors, all of a sudden, Royce had a different engineering challenge. It was a case of 'building the better mousetrap'. He calculated he could improve on the Decauville by building his own car. The car was a fairly simple construction, although technologically challenging for the period, and lent itself to readily observable improvements. Royce got board approval to build three cars at the Cooke Street location. Henry Edmunds of Glover Cable, also a director of F. H. Royce & Co., encouraged him; this was another of Royce's 'fortunate contacts' that shaped his business life for the future. Technology was developing rapidly and companies had to commit to being involved or be left behind and no longer competitive.

'Royce Petrol Motor Cars' was added to the list of products available. This was quite a gamble for a relatively small company with limited resources. However, to maintain company growth, a new venture was called for because, as mentioned earlier, cheaper, perhaps inferior, electrical products from abroad were eroding the company's market. Keeping in mind that it is sometimes easier and cheaper to improve a product than design it, Royce set about improving the Decauville. The frame tended to distort. Royce used it as the foundation and improved it by isolating the main components and providing extra frame stability by use of torsion bars and bracing. This was the beginning of the legendary Rolls-Royce smooth ride. Royce definitely had the golden touch with engineering improvements.

Improvements to bearings, engine valves, the radiator and gear shift were among the improvements of the Royce car. The 10hp engine first ran on 16 September 1903 and by 1 April 1904, the prototype experimental car was ready for the road. It had a progressive cone clutch, lubricated shaft drive joints, big silencer, and tapered bolts. Royce's aim was to achieve perfection through attention to the smallest detail. The first public demonstration of the Royce car was the same month as the Automobile Club trials meet. It was noticeably quieter than the other cars. The 'dynasty of quiet' had begun. Rolls most certainly would have seen the car perform at the trials. Henry Edmunds drove the car on the 145-mile (233-km) London to Margate Endurance Trial. Edmunds realised the potential of the car and had a motoring friend who he thought would be interested in it; he sent plans and photos of the prototype to his friend for examination. That friend was Charles Rolls.

Charles Stewart ROLLS

The Honourable Charles Stewart Rolls was from a totally different background to Royce. Charles's father, John Rolls, was a wealthy landowner in Monmouthshire, South Wales, and held various offices in the county such as Justice of the Peace, High Sheriff of Monmouthshire, and was a commissioned army officer. John Rolls, Lord Llangattock of the Hendre since 1892, had two large estates in South Wales as well as properties in London, the English capital. Lord and Lady Llangattock were independently wealthy due to the revenue from their real estate holdings and lived the life of a financially successful family.

Charles Stewart Rolls was born with a 'silver spoon in his mouth' on 27 August 1877 at the prestigious address of 35, Hill Street, Berkeley Square, London. Sir Winston Churchill's childhood home was 48 Berkeley Square and of course a nightingale sang there accordng to Vera Lynn, a hit during the Second World War. Charles was the fourth child, following two brothers and a sister, and did not have the restrictions of possible family succession that the older siblings had. He would be what we call today a 'free spirit' or slightly eccentric individual. Both his elder brothers died during the First World War, but his sister outlived him by fifty-one years. None of the four siblings had any children.

Shortly after Charles's birth the family moved to another prestigious London area, South Kensington. Although a 'city boy', apparently he was very fond of the open country surrounding his family home at Hendre, near Monmouth. During his education at Eton College, Charles developed an interest in engineering and could often be found tinkering happily with mechanical and electrical machines. This earned him the nickname 'Petrolls' among his peers. The play on words is a combination of petrol and Rolls. His fascination with engineering did not help his studies at school and he had to 'cram' to gain entry to Trinity College, Cambridge.

While studying, he was active in the sporting life of the College, cycling being one of his achievements. He joined the 'Self Propelled Traffic Association' in 1895 to campaign against the restrictions of the Locomotive Act, which stated a the four miles per hour road speed limit had to be observed and that the automobile had to be preceded by a man with a red flag to warn other road users of an approaching automobile.

The Association was successful in raising the speed limit to a dizzying 12mph. Rolls went to Paris and joined the Automobile Club of France and while there in 1896 he purchased a very powerful Peugeot Phaeton 3¾hp car. It was possibly the first car in Cambridge and his journey back to Monmouth caused a stir among the general public. Rolls was already demonstrating that he would be a pioneer in car travel over roads designed for horse-drawn vehicles. He accepted the challenge of automobile travel and became very interested and involved in the emerging automobile industry. Rolls was one of the founding members of the Automobile Club of Great Britain in 1897 and served on its committee until 1908.

Charles Rolls graduated in Mechanism and Applied Science in 1898 and was accepted as a student member of the Institution of Civil Engineers. Rolls retained his adventuresome spirit and his family's financial backing allowed him to experience many events that would not be feasible for an ordinary citizen. One adventure that particularly appealed to him was defying gravity and he took the opportunity of a balloon flight from the Crystal Palace just before the turn of the century, in 1898. Here he had found a freedom from the roads and the automobile. The interest in flight remained with him all his life and would eventually be a large factor in his life story.

Rolls' post-university work experience included engineering in the maritime and rail industries, namely steam driven yachts and the main workshops of the London & North-Western Railway. Rolls lived a conservative home life, which gave him the energy for his competitive sporting activities. His initial interest was in racing cars with his rich friends, who could also afford to take part in such an expensive sport; they could be found at every 'goggles and dust' race throughout the country. The Honourable John Scott-Montagu and Rolls were competitors in the 1899 Paris to Boulogne Race. The race was in the traditional 'city to city' format. In 1900 Rolls received the amateur award for his performance in the London to Edinburgh race in his 12hp Panhard. Rolls was all action and Royce was all thought. Rolls' delight and his feeling right at home driving cars to their limit was evident to all who drove with him. However, in his social life he could be called slightly eccentric in dress and manners. Aloof and perhaps snobbish, he would not be appreciated today by the man in the street, but he would fit right in as salesman to high society; and it would the rich buying Rolls-Royce cars in the future, not the man on the Clapham omnibus.

In 1901 Rolls once again took to the air with fellow members of the Automobile Club and an historic decision was made to form The Aero

Club of the United Kingdom. Yet again, Claude Johnson jumped at the opportunity to become the temporary Secretary of the Aero Club; Rolls was a committee member. Simultaneously, he was a member of the Aeronautical Society, which was interested in heavier-than-air unpowered and powered flight. He was not very active in ballooning as a sport for the next four years until a ballooning competition caught his imagination. Rolls had already competed in the Gordon Bennett road race in 1903 so the Bennett international ballooning competition was a natural attraction. He selected the Short Brothers, of future aviation fame, to build his balloon.

One of Rolls' friends was John Walter Edward Douglas Scott-Montagu, the 2nd Baron Montagu of Beaulieu, who shared his sporting competitive drive. Schooled at Eton, followed by competitive rowing at New College, Oxford, Scott-Montagu also worked for the railways to gain practical engineering experience. A dedicated motoring enthusiast, he was the founder and editor of *The Car Illustrated* magazine. Another sporting friend was John Moore-Brabazon, 1st Baron Brabazon of Tara, who was educated at Harrow then went to Trinity College, Cambridge, and spent the summers working for Rolls as an unpaid mechanic. In 1907 he won the Circuit des Ardennes in a Minerva automobile.

This was the first French major race to be held on a closed course rather than the traditional 'city to city' format. The first closed circuit race was the 1903 Gordon Bennett Race held in Ireland. Each year the Gordon Bennett race was held in the home country of the previous year's winner. In 1902 the race had been won by an Englishman (S. Edge) and, because it was against the law in England to close public roads, the race was held in Ireland. As a thank you to Ireland, the English cars (each country entered a team) were painted shamrock green, and that is how the famous British Racing Green colour was created. You could say these closed races were the foundation of the present-day multi-million-pound closed circuit Formula One race series.

Rolls finished fourth in the 1899 Paris to Boulogne Race, and for a competitive person with money, that was not a satisfactory outcome. He sold his 8hp Panhard to Mark Mayhew and purchased a 12hp Panhard. Lieutenant, later Lieutenant Colonel, Mark Mayhew was an automobile racing enthusiast. He inherited great wealth and owned the Battersea Flour Mills. This enabled him to indulge his passion for the sport. In 1903, during the Paris to Madrid rally, Mayhew's car lost its steering and hit a tree, fortunately without serious injury to the driver.

The purpose of selling the 8hp to Mayhew was to purchase an even faster car, the 12hp Panhard, for 1900's 1,000-mile reliability trial organised by the Automobile Club of Great Britain and Ireland. It is hard to believe today that such a trial ever existed, but this was in the early days of transition from the horse and carriage to the horseless carriage. The 'fire breathing machine' was being challenged by the tried and trusted animal and conveyance. There was general public resistance due to the noise and speed of the automobile; and we have always had horse-drawn vehicles, so why change?

Except for the publicity of the London to Brighton Run (the first one was held in 1896), the automobile was mostly seen by city dwellers. The few that were seen in the countryside were driven by wealthy pioneers dressed in outlandish clothes to protect them from the elements. These consisted of a heavy coat, hat and goggles, as initially the automobiles had no windscreens. They certainly did not endear themselves to the struggling farmworker driving his cattle down the track, now called a road, or the country folk riding their horses. Scaring everything in sight with its noise and smell, the automobile posed a question to the average person: Was this really progress? The aficionados certainly believed it was, and a successful trial would provide the proof.

The '1,000-mile trial event' was promoted by Lord Northcliffe and the organising was ably undertaken by Claude Johnson, who would subsequently have a successful career and a pivotal role with Rolls-Royce. Johnson travelled the length and breadth of the course and set up the accommodation and food for all the participants. In those years, this meant the driver and his mechanic. Johnson charted the route and urged the route's local inhabitants, indeed the whole country, to witness this armada of automobiles, which sought to establish once and for all that the horseless carriage was here to stay.

It would be a serious challenge for man and machine. Considering that most recreational motoring journeys were less than 100 miles, this 1,000-mile journey was definitely something new, a leap of imagination by the organisers. The London to Brighton Race was 54 miles long and the original run, the 1896 Emancipation Run, was held to celebrate the passing into law of the Light Locomotive Act, which raised the speed limit to 14mph and did away with the need for a person to walk in front of a mechanised vehicle waving that now comical red flag to warn other road users.

The route was through England and Scotland and featured exhibitions, hill climbs and speed trials, all designed to present the automobiles in their best light and get the citizens of Great Britain onside to support the transition to the new era. The entry list read as a veritable who's who of the 1900 motoring elite – Charles Rolls and his motor sport friends Scott-Montagu, Mark Mayhew, Claude Johnson, J Siddeley and Herbert Austin. It is recorded that C. S. Rolls had the fastest time for the Sharp Fell hill-climb in his 12hp Panhard.

Although by now an accomplished driver, we should not forget that Rolls was still an engineer. His time with the railways had given him a good engineering foundation and he was known to have submitted several patents and continued an interest in mechanical and electrical developments. At twenty-three years old, he realised that with this background, as well as a wide circle of wealthy motoring friends and his own interest in getting the next fastest automobile, there may be a business opportunity. There was a dearth of car dealerships in Great Britain and two years after the great 1,000-mile trial in 1902, he opened C. S. Rolls & Co., an importer of 'fine' French Peugeot and Belgian Minerva cars, in Fulham, London.

Located under one roof, the company featured new and previously owned cars in showrooms and, most importantly for the fledgling business, there was a fully equipped machine shop. This shop could make parts rather than wait for them to arrive from the Continent. Imported foreign cars did have their drawbacks and that also was a business opportunity, which would shortly become apparent. In 1902 Rolls did not have much time for racing anymore but did manage to enter the Paris to Vienna race. He did not finish this gruelling race over very steep terrain; his car broke down the first day. It is interesting to note that the winning car was a shaft-driven Renault K-Type car.

Rolls also continued his adventures outside the motoring community. An adventure in 1903 narrowly avoided disaster. A rapid ascent followed by a rapid descent in a balloon ended in an accident. The occupants of the balloon walked away shaken but unhurt. Again, Rolls was pushing life's boundaries in spite of having the responsibility of a growing business. He now had a location in the West End of London. He advertised in his friend John Scott-Montagu's *The Car Illustrated* magazine and C. S. Rolls & Co. continued to grow and expand into new areas. Rolls offered car loans, hire purchase, to those who were less wealthy and desired a car.

He also offered a course to train chauffeurs to drive properly and efficiently. This included the point of etiquette of not walking in front of the vehicle after assisting the passengers, which also allowed for a final check of the tied down luggage, petrol cap etc., before setting off. Not content with that, Rolls included how to maintain the vehicle. This was recently illustrated by the character Tom Branson, chauffeur in the television series *Downton Abbey*, seen working on a car as he courted Lady Sybil, the Earl and Countess of Grantham's daughter.

There were two major defining moments in the history of Rolls-Royce. The first was when Claude G. Johnson joined C. S. Rolls & Co., and the second was when Charles Rolls met Henry Royce. Johnson, known as CJ, was an extrovert well suited to organising the first automobile exhibition in England for the Imperial Institute, South Kensington. The year was 1896. Johnson was first secretary of the Royal Automobile Club (RAC) in 1900 when he organised the famous 1,000-mile trial. He left the RAC in 1903, originally for a manufacturing venture, but became joint manager with Charles Rolls of C. S. Rolls & Co., finding high quality cars for friends. Johnson was friends with Lord Northcliffe, the owner of the *Daily Mail* and *Daily Mirror*. Publicity, therefore, was not a problem for the growing company under the guidance of Rolls and Johnson.

In 1903 the British Motor Service Volunteer Corps was founded by Rolls' friend Lt-Colonel Mark Mayhew. The Corps provided cars and motorcycles for military purposes. Rolls became a volunteer with the rank of captain. With Johnson looking after the business, Rolls could get away to pursue his passion for competition. He drove his 80hp Mors to an unofficial world land speed record for the flying kilometre on the Duke of Portland's estate. The course was deemed not entirely flat and the record was not ratified. The Mors was a French-built car, its main rival being the Panhard. The Mors had been around since 1897.

Rolls took part in the 1903 Gordon Bennett Trophy Race in Ireland. Hill climbs, flying mile and Match races continued over many days and venues. Rolls had some success, winning at the Phoenix Park, Dublin, and some near successes – his friend Scott-Montagu beating him by 4 yards. Remember, these were two cars racing wheel to wheel on narrow roads with 1903 technology; not a sport for the faint-hearted. During this time, he also continued to go ballooning. Just as well he had a competent partner in Johnson to look after the business.

Although driving many varieties of French cars, Rolls always had a hankering to have his name associated with the design and development of a car, no doubt from his engineering background. It would also be a business benefit to promote a British car in opposition to all the foreign cars flooding the market. However, his initial thoughts, perhaps, were more towards a luxury car rather than an out-and-out racing car. After all, he had a wide circle of rich friends who would certainly support him and a home car industry. Did such a car exist? He did not know of any because certainly he would have sold them or at the very least would have driven one.

Rolls belonged to many organisations and he was well known within them. The 'old boys' network' was about to play a large part in his life. Did Rolls and Royce know about each other before they met? Probably they did. Royce was widely known for his quality engineering, both electrical and with cranes, and probably was known by wealthy society who used his company and its products. Rolls, a member of that wealthy society, was widely known for taking part in motor sport and for his company, which benefitted from the publicity in the newspapers. These two accomplished men, in their respective fields, just needed a catalyst to bring them together.

Rolls–Royce

That catalyst was another successful engineer, Henry Edmunds, who later became known as the 'Godfather of Rolls-Royce'. Edmunds, a director of Royce Ltd., appreciated the speed, reliability and engineering of the Royce car and was aware that Rolls was looking to associate himself with a quality British car. Both Edmunds and Rolls were members of the Automobile Club, the 'old boys' club' in question. He persuaded Rolls to make the effort to meet Royce at the Midland Hotel, Manchester, in the spring of 1904. The two men apparently took to each other immediately and agreed to form (in alphabetical order) the Rolls-Royce car company. Rolls wanted to drive and sell the best cars, preferably with the Rolls name on them, and Royce wanted to design and build the best cars. The automobile industry would never be the same.

The popular account and entertaining reading is that Rolls borrowed Royce's two-cylinder car and drove it back to London. He was so excited at his find that he awoke Johnson in the middle of the night to give him a test drive in the two-cylinder Royce car. Rolls-Royce scholars cannot find proof that this event occurred as described. The other unverified version, probably closer to the truth, is that Rolls saw the Royce car before going to Manchester. The stage was set for a magnificent partnership that would produce beautifully designed cars for more than 110 years to date. Until then Rolls had been selling Panhard, Mors, Minerva and Clement cars, all foreign. Finding a car that exceeded the best foreign cars then being produced, Rolls and Johnson were so impressed that they agreed to sell all the two, three, four and six-cylinder cars that Royce could produce. The 23 December 1904 proviso would be that the cars would be sold under the name Rolls-Royce. One meeting and Rolls had his dream.

Rolls had in mind a 15hp car for his discerning customers, so a decision was made to develop a three-cylinder version of the 10hp Royce car. This was the first Rolls-Royce badge collaboration, initially red but reputedly changed to black to avoid a colour clash, of a long line of cars. Although Rolls and Royce are long deceased, the reputation of the marque continues to this day in the automobile and aviation industries. What made the Rolls-Royce cars so different? Superior handling, reliability, all-weather comfortable driving made the cars stand out from their rivals. Good steering lock, long wheelbase and flexible chassis all added to the Rolls-Royce smooth ride.

They say that in marriage, opposites attract and this was certainly so in this business – but with the added bonus of an engineering interest common to both men. Royce enjoyed being in the background, the engineering boffin labouring to exhaustion. Meanwhile, Rolls the aristocratic adventurer was being the 'face' of Rolls-Royce. He had all the contacts – journalists, publishers and respected engineers – to promote the company cars. What a combination it turned out to be, with Ernest Claremont and Claude Johnson behind the scenes being the continuing guiding light for the business aspects of the organisation. Rolls said of Royce: 'I was fortunate enough to make the acquaintance of Mr Royce and in him I found the man I had been looking for.'

The design and build of the 15hp car was a great engineering advance not only because of the more powerful engine but also the changes to the frame and suspension. The 15hp car was to have as many common parts with the 10hp as possible, which would save further design and manufacturing costs. The track and wheelbase were both increased and, along with the suspension improvements, the cars were now taking on a reputation of their own. The engine sizes increased correspondingly with now three, four and six-cylinder versions available.

To keep these engines cool, a new and very distinctive radiator shape was designed. Architecturally it had the same shape as an ancient Greek temple portico, which by engineering design provided more efficient cooling than the previous shapes. The new radiator – the 'Pantheon' grille – with the new badge, made the Rolls-Royce car easily identifiable, and they still do. In 1906 Johnson, the marketing guru, focused Rolls-Royce on the 40/50hp six-cylinder model. More than 6,000 would be built and it became known as the Silver Ghost. Johnson has been referred to as the 'hyphen in Rolls-Royce'.

Rolls knew the value of sporting success to enhance sales and entered the 20hp car in the Isle of Man Tourist Trophy Race in 1905 and 1906.

Rolls won the 1906 race in a lightened car, the Light Twenty, with Rolls-Royce wire wheels. Eventually these wheels were replaced with the Dunlop version as standard. Johnson also displayed their new design, the 40/50, loaded with passengers at hill climbs and at speed on even surfaces. In 1906 Rolls broke the Monte Carlo to London record and would have done a much quicker time except for events beyond his control. Of course, these successes were widely reported and acclaimed by his friend Lord Northcliffe's newspapers. Rolls started increasing his ballooning adventures again in 1906 after the announcement of the Gordon Bennett ballooning competition at the end of 1905, a commitment made possible now that he had his long-sought-after British car for his customers and with Johnson so ably managing the business.

Rolls achieved his aeronaut certificate, number four in the Aero Club, in 1906 and also purchased several balloons in the same year. A leading advocate for the Aero Club, he introduced various ballooning challenges, organised competitions and prizes, and offered the family home, The Hendre, for balloon ascents. Rolls was of course aware with all his ballooning experience of the balloon's massive limitation, that it was at the mercy of the wind (although a very limiited amount of control of direction could be applied by ascending or descending to reach a level where the wind was blowing in a different direction). And of course he was also aware of what the Wright Brothers had done in 1903.

Behind the scenes, Royce was under immense pressure as the focus was now on cars and not electrical components. Balancing of the engines for smoothness, crankshaft torsional damper and strength was just one of the many new engineering challenges facing the rapidly growing company (in fact, such challeges were faced by practically the entire automobile indistry). Gearboxes now had to absorb and transmit higher torques to the rear axle as the output of the engines grew. Royce was the leading engineer and brains behind all this development – it was exhausting work and was taking a subtle toll on his health.

An example would be the relocation of the engine forward to underneath the driver's feet and away from the passengers, thus providing them with less noise, vibration and combustion smells. The body of the cars were becoming bigger and more luxurious to provide the utmost comfort for the passengers. Engine design was rapidly changing, as various combinations of bore and stroke were tried, and a new engine called the 40/50 was very successful in the 1907 Scottish Reliability Trials. Rolls-Royce established a new non-stop reliability distance that was twice the amount of Siddeley, the previous record holder. The 40/50 refers to the tax power rating, which is based on the engine cylinder dimensions. It was used to calculate payable government road tax.

Claude Johnson christened the 40/50 engine the 'Silver Ghost', which eventually became synonymous with the car itself. Johnson was the brains and drive behind the advertising campaign. He used slogans such as: 'When you see a six-cylinder car but cannot hear its Engine or Gear you may be certain it is a Rolls-Royce, Silent as a Ghost, Powerful as a Lion, Trustworthy as Time.' Johnson continued to promote Rolls-Royce with huge energy throughout his career.

Rolls-Royce Limited was registered on 15 March 1906. Ernest Claremont was elected as its first chairman. Financially it was not all plain sailing for Rolls-Royce and if Arthur Briggs had not subscribed to a failing flotation for public funds, the Rolls-Royce story would have been completely different. The company absorbed C. S. Rolls & Co., in 1907, and the car-making part of Royce Ltd., in Cooke Street. Diversity was important to the new company. Trafford Park still manufactured cranes and Rolls-Royce expanded to the USA; a bold, even reckless move as the company was now competing for sales with the industrial might of the USA auto industry, namely Ford, Cadillac, Oldsmobile, Dodge and others. Probably the thought was that Rolls-Royce was offering a superior product that had the mystique, also perhaps social cachet, of a European automobile.

Rolls went to the USA and in early 1907 exhibited Rolls-Royce cars. He attended the Association of Licenced Automotive Manufacturers show in New York and displaying his driving prowess once again, was awarded a Silver Trophy in an auto race against bigger horse-powered cars. He also met the Wright Brothers, which some sources have indicated was the real reason for the American visit.

Rolls was appointed to the technical committee of the Aero Club, which enabled him to get a flight in an airship in France in November 1907. He also had time to meet Wilbur Wright who was in France to advance a sales campaign of the brothers' aeroplane. Rolls continued to be the social 'face' of Rolls-Royce and he pressed the company's board to consider manufacturing airships. It would seem that Rolls had moved on from the car industry already in 1907, and had his mind on airborne opportunities for the company.

By 1908 Moore-Brabazon had transferred his focus to aviation and had learnt how to fly. He is credited with being the first Englishman to fly a heavier-than-air craft under its own power in 1909 on the Isle of Sheppey. Moore-Brabazon and Rolls purchased their first balloon together; built by Short brothers, it was called the *Venus*. In 1909 Moore-Brabazon and Rolls were photographed with the Wright Brothers at the Royal Aero Club where Moore-Brabazon held Royal Aero Club Aviator's Certificate number one. Rolls and his friends were mixing with the elite of the flying fraternity. Shortly thereafter his wife persuaded Moore-Brabazon to give up flying. The reason we will discover shortly.

Expansion was in the air. Ernest Hives joined Rolls-Royce as a tester at the Derby Works in 1908. The board wanted to triple the output of Cooke Street and Royce and Claude Johnson were commissioned to find new premises for car production. The site chosen was Nightingale Road in Derby. Royce oversaw the move, without loss of production, and the choice of equipment and its installation. Again, Royce became ill with the stress of the move but recovered in time for the official opening of the Derby Works on 9 July 1908. The announcement was inevitably covered by Lord Northcliffe's newspapers. Edmunds attended the Works opening but then fades from the Rolls-Royce business scene.

Rolls made the introductory address, Royce yet again remained in the background. He introduced Lord Montagu of Beaulieu by pointing out that to the pioneer motorist he really needed no introduction. Lord Montagu had continually educated the House of Commons and House of Lords on the importance of the motor car to the future of civilisation. Rolls stated that Derby was chosen because of the co-operation of the Derby Borough Development Committee. He mentioned that the Works planning and construction had proceeded slowly to ensure they got things right the first time, as per Royce's accurately laid scheme. Rolls emphasised the fact that the Rolls-Royce aim was to turn out a small number of cars to the highest quality for discerning customers.

Lord Montagu, editor of *The Car Illustrated*, siad that he was there to support his friends Mr Rolls, 'Charlie', and Mr Johnson. He drew attention to the fact that the three men had formed a very powerful consortium, Johnson the businessman, Rolls the skilled driver and Royce the mechanical genius. He noted the surrounding green fields for future growth. Lord Montagu declared that he had ordered a Rolls-Royce and did not mind that it would be a late delivery, as any car so good was worth waiting for. What an endorsement at the opening of the new Works.

Rolls visited France to make the big aeronautical leap from ballooning to learning how to fly an aeroplane. France at that time was the leading nation in Europe for aviation and Rolls planned to learn how to fly the Frères Voisin biplane. Wright secured a manufacturing agreement and one of his first customers was Rolls, who put in an order for the Wright

Flyer. A short while later, after viewing many Wright flights, he was taken on his first flight in October 1908 near Le Mans, France. Rolls had certainly dedicated himself to the world of flight with ballooning, with model gliders, and now powered heavier-than-air machines – the aeroplane. Not only that, he foresaw the use of aeroplanes both offensively, for reconnaissance, and defensively, to intercept enemy airships. Unfortunately, it took a little longer for the Government to agree; the wheels of bureaucracy grind slowly.

The dedication of Rolls strictly to Rolls-Royce and its cars was coming to an end. The company had no interest in aviation at this time and, judging by Rolls' sporadic car demonstrations, Rolls had no interest in the automotive industry. Rolls, ever the 'bon vivant' businessman, proposed that Rolls-Royce form a partnership with the Wright brothers to manufacture the Wright Flyer in Great Britain. Conservative business sense won over diversity and opportunity; the proposal was turned down by Claremont and the Rolls-Royce board. At the end of 1909, Rolls resigned his position as the Rolls-Royce technical director. The Rolls and Royce close working relationship only lasted five years but it was long enough to launch an historic automotive dynasty.

Short Brothers subsequently picked up the contract to build six Wright Flyers, one of which would be allotted to Rolls. The Wright brothers visited London in May 1909 and when Rolls complained of problems in getting delivery of a Wright Flyer, Orville Wright suggested a glider for the interim.

July 1909 was the first crossing of the English Channel by an aeroplane. Unfortunately for the British, it was by a Frenchman, Louis Bleriot. Lord Northcliffe's slogan of 'Britain is no longer an island' was (sort of) fulfilled three years after he made it. 1909 was the year that a British-designed aeroplane first took to the air, piloted by A. V. Roe.

Finally, in August 1909, Rolls successfully achieved a glider flight at Standford Hill, Eastchurch, on the Isle of Sheppey, a similar coastal location to the Wright's first flight at Kitty Hawk, North Carolina, USA. There was no stopping him now. The gliding activity was interspersed with visits to France to observe powered flying meets and exhibitions. Rolls was ready for the next big aviation step, powered flight. In October 1909, his Short's Wright aeroplane was delivered to the Isle of Sheppey.

Meanwhile, back at the office, the Rolls-Royce drawing office was tasked with drafting a detailed scale drawing of all car parts. In 1910 R. W. Harvey-Bailey joined Rolls-Royce and was ably assisted in the department by Bill Clough and by managers of the Engine, Frame, Gear and Electrical Sections. This was the beginning of a huge database, created without the help of a computer. The ever-changing parts requirements at times made some parts obsolete, while other parts could be reused in a different model. Keeping track of the inventory using lists and catalogues was a mammoth task. This does not even take into account the various materials that had to be used to produce the part. The industry was growing. In 1904 there were more than 8,000 cars registered, in 1914 there were in excess of 132,000.

Rolls succeeded in making several powered hops down a track on the hill near Eastchurch. He was assisted by a catapult using a system of a dropping weight and cable to accelerate the trolley, with the aircraft sitting on it, to flying speed. Then it was time to attempt a real flight – man, machine, power and lift. The fateful day, as it happened, was 2 October 1909. Too much elevator caused the aircraft to pitch up off the rail and subsequently stall and cause extensive damage in the resulting crash. As the old adage goes 'any landing that you walk away from is a good landing,' and Rolls had had his first of many good landings.

I suspect that obsessive would be a good adjective to apply to Rolls as regarding his interest in flight and the aviation scene. At this time of his life it is difficult to find any mention of cars in Rolls' history, probably a disappointment to Royce as he continued to labour to improve and design new cars. Rolls continued to collect qualifications, prizes and cups for his flying, longevity and distance, but he had his fair share of incidents and accidents. Finally, in January 1910 he asked the Rolls-Royce board to cut back some of his company duties so that he could devote even more time to aviation. Flying on the Isle of Sheppey, trips to France to fly where he was awarded the French Aero Club's Aviator's Certificate, constant improvements and changes to his aircraft, the sheer strength and effort required to fly the aerodynamically unbalanced aircraft – all must surely have sapped his stamina, but they certainly never dampened his enthusiasm.

In March 1910, he was awarded the second Aero Club's pilot certificate. Rolls cemented his position in the history books as one of the pioneers of aviation in the UK. The public looked on and expected that he would bring aviation leadership and glory to Great Britain. Wheeling and dealing with the military over aeroplanes also added to his repertoire. In this regard, he was a visionary. Rolls was amassing his own fleet of aircraft, including the French Wright, eighteen months after ordering it, a Short Wright, and a Sommer Farman.

In April 1910 Rolls competed in Nice. By June he was in Dover, and successfully completed a round-trip to France. This was the first two-way crossing, also non-stop, of the English Channel and the first from England to France. Rolls was awarded the Royal Aero Club's Gold Medal and became a national hero. Wrecks, rebuilds, repairs, overhaul, flying, competitions, travel with a different bed every night, and public appearances made up Rolls' life. He envisioned a Rolls aircraft company.

In July 1910 Rolls was in Bournemouth for the International Aviation Meeting. On the 12th Charles Stewart Rolls was killed in a flying accident in his French-built Wright aircraft – a very sad occasion for Rolls-Royce and Great Britain as one of her aviation pioneers succumbed to the inherent dangers of this new technology. A full account of the accident may be found in *Charles Rolls – Pioneer Aviator* by Gordon Bruce, Historical Series No. 17, Rolls-Royce Heritage Trust. Rolls was the first Englishman to die in a powered heavier-than-air machine. Aviation had robbed Royce of his business partner, his friend and 'the face' of Rolls-Royce. Many forms of memorials have appeared for Rolls over the years, such as stained-glass windows, statues, plaques and his name on a British European Airways aircraft.

Life goes on

Charles Sykes worked as an illustrator for John Montagu, editor of *The Car Illustrated*, and knew Johnson of Rolls-Royce through him. Sykes was commissioned to illustrate the December 1910 Rolls-Royce catalogue. It was so successful that Johnson commissioned Sykes to design the Rolls-Royce company mascot. Johnson wanted the mascot to reflect the Rolls-Royce status and to be set apart artistically from the mascots of other manufacturers. The result was 'The Spirit of Ecstasy', more commonly known as 'The Flying Lady'. The ethereal figure predominant on the Rolls-Royce car's bonnet represented the speed, energy and silence of the car itself.

In 1911 a memorial statue to Rolls was unveiled at the Shire Hall in Agincourt, Monmouth. It is a bronze figure, of Rolls examining an aeroplane, atop a pink granite plinth. The inscription reads:

Erected by public subscription to the memory of the Honourable Charles Stewart Rolls, third son of Lord and Lady Llangattock, as a tribute of admiration for his great achievements in motoring, ballooning and aviation. He was a pioneer in both scientific and practical motoring and aviation and the first to fly across the channel from England to France and back without landing. He lost his life by the wrecking of his aeroplane at Bournemouth July 12, 1910. His death caused worldwide regret and deep national sorrow.

With a no-racing policy in effect, Royce was now putting all his energy into the 40/50, which was undergoing developmental changes, such as modified rear elliptical springs. His health continued to fail. In 1911 he was hospitalised and upon release a private nurse, Ethel Aubin, was hired full-time to care for him. Johnson was concerned that the future of Rolls-Royce would be grim without the genius of Royce. The proposed health solution was for Royce to spend the winters at Le Canadel, France, and the summers in the south of England. This was to ensure that he was in a warmer climate and would have a more relaxed 'home environment' work routine.

Would it work? This was before the ease of communication by email, so any communication with him would require letters, parcels, telegrams, personal visits and phone calls. Although removed from the environment of Rolls-Royce, ensuring his remotely generated ideas and designs were actioned correctly was an added stress. When he departed to France with his nurse, leaving his family in England, a rift was created that was never reconciled.

Le Canadel was established in 1912 as a colony of Rolls-Royce personnel. Claude Johnson, Royce, the design office and the designers all enjoyed the hillside location with its good weather and sea view. Harold Nockolds emphasised in *The Magic of a Name* that 'His mind undistracted ... Royce kept his staff busy with a continual stream of ideas from his fertile brain.' As it subsequently turned out, Royce would be separated from 'the Works' for the rest of his life.

How did Rolls-Royce survive after 1910 with its founding members either deceased or ill and requiring special treatment? Obviously, there were some very special and capable people guiding the company. Initially, it was Ernest Claremont and Claude Johnson. They would be joined by other equally dedicated and capable people as time went on. The cars had to be manufactured and, more importantly, sold, regardless of personnel issues, to keep Rolls-Royce viable.

In 1912 a memorial statue was unveiled to Rolls in Dover, Kent. The inscription reads:

CHARLES STEWART ROLLS the first man to cross the Channel and return in a single flight June 2, 1910

The year 1912 was not a good one for Rolls-Royce as their entered car failed to climb the Katschberg Pass during the Austria Automobile Club event. Royce attacked the problem and all four Rolls-Royce cars swept the field in 1913. Charles Freestone wrote in *Autocar* magazine '…the sight of a group of [Rolls-Royces] running up a mountain road at high speed … it was a spectacle indeed.'

Royce, despite his infirmities, was the chief designer who came up with the innovative designs and delegated these ideas to the design team, who would report back to him on a regular basis. The design team were expected to change their work as per Royce's observations, there was only one chief designer and his word was law. He was not an easy person to work with but Royce did have a successful track record. The designers had to keep their own egos in check. The completed design submission would be sent to the Derby Works for implementation. The final filing of all these engineering changes would be done at West Wittering, Royce's English summer home.

Royce certainly had an eye for picturesque locations in which to work and live. West Wittering was in keeping with Villa Mimosa in Le Canadel, France, and St Margaret's Bay in England. All three were beautifully situated coastal homes in quiet areas. West Wittering was a small village where Royce's home, Elmstead, was one of the few houses in the village at that time with its own power source, hence its own light and heat. In 1921 the designers were still using oil lamps for light in the rented Camacha studio, which was ironic considering Royce's beginning in the electrical business. Pencils, adjustable set squares and erasers were the tools of the trade in those years.

However, the dark clouds of the First World War were looming, and Royce moved back to England in 1914. Royce and Nurse Aubin had to move westwards to West Wittering from their original location, St Margaret's Bay near Dover in Kent, to avoid possible cross-Channel artillery shelling. James Radley, on a visit to Royce, had pleaded with him to install a 40/50 engine in an airship of the Royal Naval Air Service (RNAS) as it was presently underpowered. Radley's request was denied. Radley was a reserve officer in the RNAS and an accomplished aviator and car racing driver. In 1912 Radley had entered a Rolls-Royce Silver Ghost into the 1912 Austrian Alpine Trial and he also competed in 1913 and 1914. The 1913 three-car victorious trial team was ably managed by Hives. He assisted Royce in solving the problem of the rough judder in the servo-assisted four-wheel brake system, which had superseded the transmission brake system.

D. Napier & Son Limited was a British automobile and engine manufacturer, a competitor of Rolls-Royce and one of the most important aero engine manufacturers in the inter-war period. Napier was founded in 1808 and one of its first designs was a steam-powered printing press. Montagu Napier assumed control of the company in 1895 at age twenty-five. He teamed up with Selwyn F. Edge in the same relationship as Royce and Rolls, Napier was the engineer and Edge was the driver/salesman. Edge's Motor Power Company took the entire output of the Napier automobile factory. A dispute in 1912 resulted in Napier taking over the entire distribution and sales company.

Edge was a racing peer of Rolls and they knew each other well, both as competitors and teammates. In 1900 Edge, with Rolls as riding mechanic remarkably enough, failed to finish the Paris-Toulouse-Paris race in a 16hp Napier. The Napier took part in the Gordon Bennett races, in which Rolls also competed. A move in 1903 to Acton, West London, continued the Napier production, employing 1,200 people by 1907. Napier cars continued to accrue awards in various categories up to the 1908 Tourist trophy race. Early in the First World War Napier was contracted to build aero engines for the war effort. Similarly to Rolls-Royce, Napier redesigned his own very successful W block 12-cylinder Lion engine in 1916. This engine went on to establish land speed records. Trucks, ambulances and the Sopwith Snipe were manufactured by the Napier Company for the war effort. The last Napier cars were made in 1924. An attempt was made in 1931 to purchase the bankrupt W. O. Bentley Company. Napier was out-bid by none other than Rolls-Royce.

In 1914 Royce's design team consisted of Albert Elliott and Maurice Olley, who worked extremely well together, another example of 'opposites attract'. Elliott remained with Royce, only moving to Derby in 1931 just before his death. Olley left England in 1917 to be Chief Engineer at Rolls-Royce of America in Springfield, Massachusetts. He subsequently left Rolls-Royce in 1930 for the Cadillac car group, forced out by the world economic collapse's effect on Rolls-Royce.

The start of the First World War forced some changes onto the Rolls-Royce Company. Planned cutbacks were cancelled, and some 40/50 cars were converted to armoured cars. A most important historic change, for Great Britain and the freedom of the world, occurred when the Government of Great Britain & Ireland asked Rolls-Royce to manufacture the Renault V8 and the RAF 1a aero engines. In true Royce fashion, the company's founder criticised the design of the engines and indicated that he would design and produce a better engine for the government. That government request sowed the engineering seeds that would eventually grow and mature into the great Merlin engine.

Now there was feverish activity because of the war and Royce threw himself once more into the designing of a suitable engine. Royce started with a 200hp version. The Rolls-Royce Works at Derby had a collection of other manufacturer's engines to be studied and improved upon and this included a German Mercedes engine. Hives set up a development testing division. The result was the first Rolls-Royce aero engine, the liquid-cooled V12 named the Eagle. The Eagle engine first flew in a Handley Page 0/100 bomber on 18 December 1915.

Other aero engines were produced during the First World War: the 285hp Falcon, the 105hp Hawk and the gigantic 650hp Condor at war's end. During the inter-war years, between the First and Second World wars, the aero engines were the Kestrel, Goshawk, Peregrine, Vulture, Buzzard, R-engine and the crowning glory – the PV12, later known as the Merlin.

These early days of aviation were filled with daring aviators and facilitating benefactors who both, in their own way, contributed to the advancement of the science of flight. One of these benefactors, a close friend of Rolls, was Lord Northcliffe, who through his newspaper the *Daily Mail* provided large sums of money for aerial achievements. An example would be the £1,000 prize for the first crossing of the English Channel, won by Louis Blériot in 1909. In 1913 Northcliffe offered £10,000 for the first crossing of the Atlantic. The offer was withdrawn during the war and a modified competition was reinstated in 1918.

Among the many manufacturers – all but one using the Rolls-Royce Eagle engine – attempting the first non-stop Atlantic challenge in 1919 was the Vickers Company using a modified Rolls-Royce Eagle-powered Vickers Vimy bomber. Rolls-Royce provided the engines for six of the seven entrants. The Vimy, with pilot John Alcock and navigator Arthur Whitten Brown, departed St John's, Newfoundland, Canada, on 4 June 1919 and nearly sixteen hours later crash-landed in a bog near Clifden, County Galway, Ireland, having flown 1,890 miles. Rolls-Royce had yet again confirmed their reputation for aero engine reliability. Alcock and Brown were both knighted by King George V at Windsor Castle for their historic flight, Sir John and Sir Arthur, and truly became 'Knights of the Air'.

Notwithstanding the wartime need to produce engines as quickly as possible, Rolls-Royce had no wish to downgrade its hard-earned reputation for the quality of its products. Like any business, Rolls-Royce had to decide where it was going to place its working capital, in factory infrastructure or new, more modern tools. Added pressure was coming from the Government who wanted Rolls-Royce to produce the RAF engine in addition to their own successful line of Rolls-Royce engines. Down-tooling and up-tooling for the different engines was detrimental to production. Rolls-Royce suggested that someone else make the RAF engine and that the types of engines manufactured be restricted to those that had proven themselves to be successful.

Johnson considered producing in the USA but was loath to divulge company production secrets, which could harm Rolls-Royce after hostilities ceased. Competition would resume, and Johnson had a responsibility to the shareholders and investors to continue providing profits. There were perhaps two positive things that came from the failed visit to the USA. The name Rolls-Royce was given exposure across the pond and, more importantly, Rolls-Royce had technically contributed to the design of the epicyclic gear for the American Liberty engine. Rolls-Royce was proving its worth to the world in the aviation industry, in addition to its reputation in automobile production.

Following Armistice Day on 11 November 1918, Rolls-Royce realised that the demand for aero engines would decrease and that the marketing and engineering focus should revert to the car industry. The 40/50 had been successful both as a staff car, ambulance and an armoured car. The armoured car was used against the Turks at Gallipoli, Germans in West Africa, in Mesopotamia and most famously of all by T. E. Lawrence, 'Lawrence of Arabia' who said that 'A Rolls in the desert was above (worth more than) rubies'. The 40/50 was also used as transport for military commanders, French politicians and King George V and Queen Mary on their visit to the Western Front, where it had developed a reputation for reliability. It was quite costly to operate owing to its size, and there was an initial thought about perhaps designing a smaller vehicle in the future.

During the war, Royce worked primarily on aero engine design but still thought about future production improvements for the automobile. Royce decided to install an electric starter motor when car production resumed. Immediately post-war, there was a mini-boom caused by pre-war back orders, which was followed quickly by a general slump in the economy due to increased competition among many companies for a very limited market. In 1923 ninety companies produced approximately 36,000 vehicles, an average of 400 per company. Costs, selling prices and road tax all increased, with the added financial burden of sales commissions and increased valuation of the pound soon causing the purchase price to decrease, with slim profit margins resulting. Something, the product or the way of doing business, had to change – and quickly – for Rolls-Royce to remain viable.

One of the changes was that in 1922 a 3-litre car, the 'Twenty', was introduced with a resulting price and operating cost saving to the customer who wished for a Rolls-Royce car. It was aimed at a different segment of the market, the owner-operator professional. Not only had the car to be at an attractive price but it would have to perform to specifications and be reliable to boost sales. It proved problematic, with bodies too heavy for the lower power. The post-war economy was not performing well, unemployment was up while sales went down. Lord Northcliffe, ever a friend of Rolls-Royce, suggested to Johnson that Rolls-Royce needed to advertise more with testimonials from well-known public figures. Rolls-Royce was located in three main areas: design wherever Royce was, West Wittering or Le Canadel; directors and sales were in London; and production in Derby. Derby did have a small design centre run by R. W. H. Bailey, which produced the Falcon, Buzzard and R aero engines.

In 1923 a memorial statue was unveiled to Royce in the Derby Arboretum. Two subsequent moves, in 1972 and 1990, relocated the statue to the home of Rolls-Royce in Moor Lane, Derby. The inscription reads:

F. Henry Royce. Engineer, born 1863, at Atwalton near Peterborough. Owing to misfortunes in childhood he was almost entirely self-educated. His work included the design and production of the Rolls-Royce motor cars of world-wide reputation, which were used for most important work in the Great War and also the design of the Rolls-Royce aero engines of which a greater horse power was employed by the Allies than of any other design. Aeroplanes with these engines made the first direct flight across the Atlantic in 16 hours and the first flight from England to Australia. This statue was erected by the shareholders of Rolls-Royce Limited in 1923 at which date he was still serving the company as Engineer-in-Chief.

In 1924 Roy Robotham had set up a testing system in France at Chateauroux. The cars were tested for endurance in 10,000-mile blocks; new models may require three blocks prior to release to the public. The resulting required changes would be agreed to by a consortium of department heads and of course the managing director, Claremont.

Three years later, in 1925, a larger version of the Ghost engine was released. At more than 7.5 litres capacity the New Phantom, later called Phantom I, coped better with the ever more sophisticated and heavier bodies demanded by those who could afford a Rolls-Royce. A four-speed gearbox and four-wheel brakes made the car suitable for city driving.

However, despite this rapid growth and success of Rolls-Royce we must remember that this is still a company beset with challenges concerning both personnel and engineering. Ronald W. Harker had joined Rolls-Royce as a 'premium apprentice' in 1925 and remembered that 'there were two hooters known as "bulls"; one blew at ten minutes to eight, the second at five to eight, then the gates closed and if one didn't get there in time, one was locked out until after lunch. If locked out, a report would be sent to the Works Manager and the sponsor of the premium apprenticeship.' A fee of £400 would buy this special Rolls-Royce apprenticeship for four years. Some workers did harbour some resentment of the premium apprenticeship programme.

Then catastrophe struck: Claude Johnson died in April 1926, and there was a national General Strike. Johnson had been the guiding light of Rolls-Royce as he took the genius of Royce and turned it into a commercial success. He was an organiser (think back to the automobile club events) and, like Rolls, was highly visible for Rolls-Royce. Making up for Royce's lack of interpersonal skills, he encouraged and rewarded the workers for their efforts to maintain standards through awards and benefits such as medical coverage and encouraging sports. Johnson was the buffer between Royce and the Works, otherwise Royce's attention to detail would have destroyed the orderly production of goods.

Changes in the form of new engineering designs to engine capacity, engine mounting, chassis lubrication and front and rear suspension continued the development as Rolls-Royce fought to maintain its lead in industry innovation for the ultimate faster and smoother ride. In 1929 the 20/25 replaced the Twenty to be itself replaced by the 25/30 in 1936. The 1929 Depression in the USA made it too expensive to retool for the new chassis as Rolls-Royce was facing diminished sales. There was also the American attitude of 'I want it now' to contend with. They didn't want to be told that the vehicle would be hand-built for the customer personally and would take eighteen months to complete. A pretty unmissable contributing factor to dropping sales was that a Rolls-Royce car was at least three times more expensive than an American car. In 1931 Springfield closed, the end of an era and, simultaneously, a new era started with the acquisition of the rival Bentley car company, which was also a victim of the 1929 economic meltdown. Bentley were noted for their racing cars, first place at the Le Mans 24 Hour Race 1927–1930, as well as an 8-litre challenge to Rolls-Royce in the luxury car market. Rolls-Royce took over the assets of Bentley Motors (1919) Ltd and formed a subsidiary, Bentley Motors (1931) Ltd.

In addition to the acquisition of Bentley, the year was a triumph for the aero engine division. Jacques Schneider, son of a wealthy French armaments manufacturer, was an aviation enthusiast who had a firm belief in the civilian and military importance of floatplanes and flying boats. Water existed everywhere, and he believed that this gave access to any country via seas, lakes and rivers. In 1912 he declared that he would give a trophy and money annually to the manufacturer of the world's fastest airplane with the requirement that it could land on water. It became the most prestigious annual competition and later was known as the Schneider Trophy Race.

The crowning glory of the competition would be that the country which won it three times in a row would keep the Trophy permanently – national pride was now at stake. The first race was held at Monaco on 16 April 1913 and won by France, so the trophy took up residence at the Aero Club of France. In 1914 the trophy moved to the Royal Aero Club. It is interesting to note that the 1914 winner, the Sopwith Tabloid, evolved into the Sopwith succession of fighter aircraft, the Pup, Triplane, Camel, Dolphin and Snipe. Supermarine would follow suit some years later. The First World War postponed the next race until 10 September 1919. Another rule was that the race-winning country would hold the next year's event.

Italy, the USA and the UK traded the trophy back and forth and the competitions in 1927 and 1929 were both won by the British Supermarine, S.5 and S.6 respectively. The 1929 Trophy was won by a Rolls-Royce powered S.6 using the R engine (more on this engine in Chapter 2). The 1928 race was cancelled due to the death of the patron, Jacques Schneider. The 1931 race was the chance for Britain to keep the trophy in perpetuity. However, the government showed initial reluctance to fund the race aircraft, and there was some doubt if Supermarine could have the S.6B ready. Rolls-Royce had done their bit by having the R engine developed to 2,350hp.

The 'brains' behind Supermarine was their Technical Director, Aeronautical Engineer Reginald J. Mitchell, a brilliant designer, who was responsible for twenty-four Supermarine airplanes. They were a collection of seaplanes culminating in his magnificent achievement, the Supermarine

Spitfire fighter. Unfortunately, like Royce, he did not get to see his contribution to the Allies winning the Second World War. He died in 1937 having seen the prototype Spitfire fly during its test flights: 'Spitfire was just the sort of bloody silly name they would choose.'

Lady Lucy Houston, a philanthropist, stepped in with a large donation and said, with political intent: 'Every true Briton would rather sell his last shirt than admit that England could not afford to defend herself.' The S.6B, with Flt. Lt. John Boothman as pilot, won the 1931 event at an average speed of 340.08mph, establishing the S.6B as the fastest aircraft in the world. The Schneider trophy remained in Britain and is proudly displayed in the Science Museum, London, as a key artefact of Britain's aviation history.

It was becoming politically apparent in the 1930s that the inter-war years were coming to an end. The Air Ministry issued Specification F. 7/30 in 1930 for a new fighter aircraft and Supermarine, among others, responded with the Type 224. It was becoming aerodynamically apparent that there had to be a better alternative to the biplane. Another war was looming, and Britain was ill equipped.

The RAF had thirty-five squadrons of biplanes, and design advances were now changing to monoplane aircraft. Simplistically, the monoplane, with half the lift of its predecessor, would require more speed which, in turn, would require more power. The evolution of the monoplane fighter and bomber aircraft had started. Royce foresaw this and in 1932, 'out of the ashes of his sickness the PV12 (Merlin) arose.' Royce defined this power unit prior to his death as a liquid-cooled 60 degrees V12 with minimum frontal area. It was one of Royce's last major decisions and what a legacy for him to leave behind, a world-renowned company with his name on it and, what would become the most famous piston engine of the Second World War.

Improvements were accelerating at a frantic pace as rival car companies competed to retain loyal customers and attract new ones with their own brand of innovations. Superchargers, independent front suspension, every conceivable combination of bore and stroke for an engine, hypoid axles, suspension bracing and links, changes to wheelbase and track, four-wheel brakes, new geometry of chassis and suspensions were all being designed, tested, approved or rejected.

It was a very busy time, not least for Royce as the 'chief engineer and bottle washer'. In 1933 a 3.7-litre engine scaled-down 20/25, based on the Rolls-Royce experimental project 'Peregrine', was badged as a Rolls-Bentley.

Sir Frederick Henry Royce, 1st Baronet of Seaton, died on 22 April 1933 at West Wittering without seeing the production models of either the Peregrine car or the first run of the PV12 aero engine on 15 October 1933. Royce had received his baronetcy in recognition of his work in helping Great Britain to win the Schneider Trophy in 1930. In 1933, a forward-looking Winston Churchill MP observed: 'Germany is already well on her way to become ... incomparably the most heavily armed nation in the world...' Can you imagine the Rolls-Royce Company scene with both founders deceased, the possibility of another war looming and the company's driving force no longer at the helm? Royce had driven the personnel around him mercilessly with his obsession, which became the Rolls-Royce tradition, for perfection. It was nothing that he was not doing to himself and he finally paid the ultimate price when his obsession destroyed his body with overwork. What a price to pay for perfection! There is a message in this for all of us.

The Rolls-Royce Company that the two founders left had a reputation for integrity, together with hard-working employees, which resulted in a higher than average quality of product; a close relationship developed between the workers and the middle and upper management and it was one big happy family in those years with many third-generation workers. A team spirit, encouraged by Rolls-Royce Chairman Ernest Hives, developed into brand loyalty for the worker and fostered good customer relations. People were proud to say that they worked for Rolls-Royce and anticipated spending their working lives with the company until retirement.

Rolls-Royce was producing, testing and developing aero engines in the early 1930s but did not have a dedicated testing facility in Derby. The solution was to establish, in 1934, a flight test establishment nearby, 20 miles from the factory at RAF Hucknall, Nottingham Airport. The test facility had an installation department to install the Rolls-Royce aero engines and a flight test department to get 'real life' information on their airborne performance. Hives is reputed to have told his testers that they were flying testers and not part of the spotted scarf brigade! Hucknall consisted of a wind tunnel, drawing and design offices, machine shop, test rigs, cold chambers and a propeller hangar.

Hucknall most importantly had a service department. This was the connection between the operator, the RAF, and the manufacturer, Rolls-Royce. Squadrons were lectured on engine operating procedures and RAF repesentatives were given tours of the Merlin assembly line. The variance in fuel consumption by pilots was a concern and could only be improved by education. Complaints were also handled by the service department. In this way, a large amount of operational data was gathered to improve future modifications of the Merlin. Not only did Rolls-Royce run an Engine Handling School at Derby on how to get maximum efficiency but company personnel were dispatched around the world to teach the correct handling of the Merlin. Not all the engine improvements were by Rolls-Royce. For example, the Schilling restrictor was designed by an engineer at the Royal Aircraft Establishment (RAE), Farnborough.

Another RAF liaison that Harker, former premium apprentice and now test pilot at Hucknall, was associated with was the Air Fighting Development Unit (AFDU). The Unit was tasked with developing tactics to gain air superiority and improve fighter aircraft handling capability. Rolls-Royce was directly involved with both aspects as the Merlin went through many changes. Improved engines were flown to the AFDU for their opinion. The Aeroplane and Armament Experimental Establishment, Boscombe Down, Wiltshire, initially tested the handling and performance of all models.

Rolls-Royce aero engine division started successfully with the First World War engine, the Eagle, and the question now was, how much would Rolls-Royce contribute to the Second World War with another bird of prey, the Merlin?

2

The Merlin Pedigree
1904–33

To the observer, the electric motor and liquid-cooled internal combustion engine were two unique items, but to an engineer they presented distinct engineering design and production problems to be solved to make them work efficiently. To achieve near perfection in both genres would take an engineering genius. Royce was such a man. We previously read how his dedication, perhaps obsession would be a better word, to achieve engineering excellence came at the cost of the loss of his health. Royce started out to improve on all that had gone before, from all sources, and he succeeded with many designs, but his crowning glory must be the Merlin. He set the engine's engineering design foundation and other great men developed and brought his ideas to production. The sad part of the story is that due to his premature death, he never heard the beautiful sound of the twelve cylinders of his creation and never saw how the Merlin played an important part in changing world history.

The Boeing 787 Dreamliner has been in the news the last few years as it represents another new generation of passenger aircraft, which is being acquired by the global airlines. It is available with engines by two manufacturers, Rolls-Royce and General Electric (GE), and has a unique feature in that the engines are interchangeable, simply change the engine pylon and install the other manufacturer's unit. Rolls-Royce also manufactures engines for the rival Airbus Industries. Clearly, Rolls-Royce is a viable engine manufacturer in the twenty-first century after a hundred years of experience designing and building engines from the Eagle to the Trent series of engines. There were of course marine and industrial engines to complement the aviation series of piston and turbojet engines, quite a history, quite a story.

The Boeing 787 coincidently had teething problems, not uncommon among new aircraft, involving its lithium-ion batteries. It was with batteries, used in home electric bell sets, that Royce got started in his engineering company. This progressed rapidly from small electrical parts eventually to designing large electrical motors used in Royce winches and cranes. In 2016 Rolls-Royce signed an £8 million contract to supply the mooring system for the world's largest semi-submersible crane vessel (SSCV), *Sleipnir*. It comprises a twelve-point mooring system including electrically driven winches, fairleads, wire sheaves and a control system.

Possibly from his time with the railways, Royce was fascinated with the engineering aspects of propulsion and movement. Royce's first motor vehicle was a water-cooled De Dion Bouton powered quadricycle. He graduated to a twin-cylinder Decauville, which he immediately dismantled and set about improving with the result that the first Royce-built and designed car ran on the road for the first time in 1904. This initial car was quickly followed by three-, four- and six-cylinder models with corresponding increases in horsepower. The Royce car had now become a Rolls-Royce car and could be seen at hill climbs, trials, Alpine trials, runs, the TT Races, and record-breaking driving and sporting events.

At the same time, prior to the First World War, other companies were manufacturing aero engines, so Rolls-Royce had some catching up to do. The most famous engine was the Wright Flyer engine because it propelled the first successful powered flight in 1903. The Curtiss Aeroplane and Motor Corporation in the USA produced a four-cylinder water-cooled inline engine. However, it was France leading the way in selection of available engines. The Antoinette engine by Levavasseur was followed by an eight-cylinder 90-degree water-cooled V engine. An Anzani engine, a three-cylinder air-cooled engine, powered Louis Blériot's monoplane across the English Channel. The Seguin brothers developed a long line of Gnome rotary air-cooled engines, the crankshaft being fixed to the aeroplane and the cylinders and crankcase rotating about it. Le Rhône and Clerget were other rotary air-cooled engines developed before the First World War.

In a short eight years, Rolls-Royce had become a household name with a reputation for reliability and innovation. Prior to 1914 the British Government relied upon the French aviation industry for aircraft supplies, which became a competition between the Royal Flying Corps (RFC) and the Royal Naval Air Service (RNAS) to satisfy their requirements. This situation was not efficient for either party and the British Government realised that it was time to develop its own home aviation industry.

A small number of aviation companies existed in scattered locations and, more importantly, these companies only had aero engines in the experimental stage prior to the outbreak of First World War. The outbreak of hostilities on 4 August caused the Government War Office (WO) to approach Rolls-Royce about contributing to the war effort by building aero engines. Royce had twice before refused to have Rolls-Royce build any aero engines and this was no exception. This time the company had been asked to build the Renault V-8 and the RAF 1a engine. Royce disapproved of both engines, which incidentally were air-cooled, but agreed to build some, eventually 220 and 100 respectively, to placate the Air Ministry. Royce's engine experience to date was water-cooled engines and he applied this knowledge to the aero engine.

Royce decided to design and produce a Rolls-Royce branded engine to the exacting standards of Rolls-Royce and not to be associated with manufacturing other company's engines. Perhaps Royce's head overruled the messages of capitulation comng from his body once again. He was still recuperating from a major operation at the end of 1912, but he instantly launched into a major project as part of the war effort. What factors precipitated Royce's change of heart to move in to an entirely new engine genre? The challenges of taking an internal combustion automobile engine and adapting, changing and developing it for use as an aero engine were enormous. The board under risk-averse Ernest Claremont turned down the request to build aero engines at an emergency meeting on 7 August 1914. A story about an airship making poor headway against a headwind was related to Royce at the time by Olley. This may have planted the seeds of the need for a more powerful aero engine, a design challenge for Royce. A further request by the WO changed the board's decision on 19 August 1914 and they agreed to build aero engines. This certainly was influenced by the fact that the French government would be taking all the output of the Renault factory – who would build the replacements in Britain? It should also be noted that none of this would have been possible without the lending institutions extending overdrafts for war work.

The main thrust of this chapter will be to analyse the Rolls-Royce collection of piston engines as they were developed to meet the ever-demanding aviation industry. The competitive engines will also be mentioned as other manufacturers vied for production orders. There will be some technical information to demonstrate how the engines were developing in generating more horsepower and engineering sophistication. The Merlin will, of course, be mentioned but the full history will be left for succeeding chapters. The overlap of the beginning of the jet engine era and the demise of the piston era is an interesting time in Rolls-Royce history as the company adjusted to the radically different technology. A detailed post-Merlin era will be left to other publications but there will be a general synopsis in Appendix I as Rolls-Royce continued to develop its jet engines successfully into the 21st century.

This is a good time to define some commonly used piston engine terms:

Cylinder	Tubular section in which the piston works
Bore	Diameter of cylinder
Piston	Cylindrical piece moving up and down in cylinder
Stroke	Distance the piston moves up and down in the cylinder
Cylinder head	Caps top of cylinder, contains inlet and exhaust valves
Valve	Device in cylinder head that regulates flow in and out of cylinder
Camshaft	Rotating shaft with special protruding shapes, the cams
Rocker	Transmits the rotational movement of the camshaft cams to operate the valves in sequence
Horsepower (hp)	Unit of measurement of power, rate at which work is done
Engine Capacity	Total swept volume of the pistons inside all the cylinders.

Author's note – all engine facts and statistics should be taken as only reasonably accurate due to the variance of test data from different sources, horsepower varies as per supercharging and altitude, various engine series (Marks), typographical and other errors of research information available. The intention is to give a relative view of the specific engine in comparison to other engines in the context of a general understanding.

Rolls-Royce Eagle

The government requirement was for three engines, a 100, 150 and 200hp version. The dissipation of heat from the cast-iron components was always a problem and Royce insisted on water-cooled engines, which was the Rolls-Royce expertise and experience at the time. Royce would certainly have examined the air-cooled Renault 80hp engine that the company had agreed to build. Royce decided on the 200hp engine, which was required for a proposed heavy bomber. Part of his research into engine design was a Mercedes DF80 six-cylinder engine that Hives had dismantled and examined, with the typical Royce approach to improving his engine design to maximum efficiency. Technical knowledge from all sources was considered to build the best engine that Royce could design. He worked closely with Albert Elliott, his principal design assistant, and Maurice Olley, initially hired as a jig and tool designer. Also part of this team was Claude Johnson at the Rolls-Royce head office in Conduit Street, London, and Arthur Wormald, General Works Manager, and Robert Harvey-Bailey, Technical Production Engineer, at the Derby Works. This working arrangement was further complicated by the fact that Royce was working from home in St Margaret's Bay, Kent, more than 300 kilometres away from the factory in Derby. This was at a time nearly sixty years before the invention of email, so the exchange of ideas and plans was done by hard copy transported by the Royal Mail or Rolls-Royce's own resources. The exigencies of wartime demanded immediate action, which dictated that there was no time to 'reinvent the wheel' but to use and develop available company resources.

Royce used the successful 40/50hp engine of the Silver Ghost car as the basis of the aero engine with engineering consideration for the top end from Mercedes and Renault for the connecting rods. The crankcase was aluminium for weight saving and strength and the other parts, such as the lubricating system, the crankshaft and piston cooling, were all state-of-the-art components for 1914 and so were included in the Eagle's design. The stroke of the piston was increased but it required the addition of six cylinders to attempt to achieve the desired 200hp required by the War Office. The final engine would look nothing like the 40/50 engine but without its technology as a starting point, it is doubtful that Rolls-Royce would have achieved such quick success with the Eagle engine. To get the required minimum horsepower Royce had to increase the rpm of the engine, which caused further engineering changes. Two epicyclic gears are mounted so that the centre of one gear revolves around the centre of the other. Reduction propeller gearing was necessary to keep the propeller rpm below the desired 1,100 rpm as mandated by the RNAS to maintain propeller efficiency. This gearing produced less strain on the crankcase than spur gearing. Olley worked on the gearing to get the reduction ratio correct. A major change was replacing the side valves with overhead valves driven by a single camshaft (SOHC).

The final design was a 60-degree Vee shaped, twelve cylinder (V-12) four-stroke engine with two banks of six cylinders and a capacity of

20.3 litres. Each cylinder had two valves, one inlet and one exhaust, and each six-cylinder bank was operated by a single overhead camshaft and rockers. Cooling and weight were always an issue and Royce decided on pressed-steel water jackets for each cylinder separately. The later engines had two spark plugs per cylinder served initially by four six-point magnetos and finally by two twelve-point magnetos. The cooling capacity was 3.1 gallons and Rolls-Royce literature stated that the oil consumption of the pressure lubricating system was 1.0 gallon per hour, very high by today's standards. The Claudel-Hobson carburettors could be controlled from the cockpit, adjusting for increasing altitude, rich mixture starting and economical cruise.

Drawings, notes and parts were sent back and forth by daily posts, by telegrams and by visits to Royce to ensure they met with his approval and would function as envisioned. It was indicative of Royce's effort that the first drawing for the Eagle engine was produced on 3 September 1914, titled *200hp Aero Engine Crankshaft*. Johnson had the notes preserved for history by creating a book called *The First Aero Engines made by Rolls-Royce*, which became known as 'the Rolls-Royce Bible'. There was no time for mistakes as the war made production urgent. By February 1915, Hives had a completed engine on the test bed and the target horsepower was exceeded at 1,600 rpm. What remained was to improve the reliability while reducing weight and fuel consumption, a typical challenge for all piston engines. The Eagle and its parts were often tested close to destruction by endurance trials as various parts and lubricating systems were tried and evaluated. Royce dictated that parts must be made to a high standard and be interchangeable, which necessitated a 1915 'quality control' long before the International Organisation for Standardisation of this century.

By January 1915 orders were being negotiated and later signed, such as the twenty-five engine order for twelve Handley Page 0/100 twin-engine patrol bombers, with one spare engine. The Eagle engines were manufactured as left-hand and right-hand units as the respective propellers rotated in opposite directions. The orders were received while the Eagle engine was still on the drawing board. Was this faith in Rolls-Royce engineering or the exigencies of war? I would like to think that it was the Rolls-Royce reputation. By March 1915 the Eagle had been bench-tested and by 4 March tested at 225hp at 1,600 rpm. This was achieved in just less than six months from engineering vision to reality – some effort from all involved. On 17 December 1915, the Eagle had its first flight in a Handley Page 0/100 bomber at Hendon, London;

this was an historic first flight of a long line of Rolls-Royce engines produced from then until the present day. Six months later, the RAF FE2d fighter was the first aircraft to enter active service with an Eagle engine. The 0/100 and Airco DH4 light bomber followed in March 1917. This horsepower was above the contracted amount of 200hp. Now the work began, and never actually stopped, on modifications to get more horsepower, reliability, lower specific fuel consumption and other improvements to make the Eagle engine a vital contributing factor to the war effort. By December 1915 a further order of 300 Eagles was placed with more orders throughout the war.

In Britain, the engine continued to improve and reached 260hp in May 1916, 296hp in December, 350hp by September 1917 and 360hp (normal rpm) in February 1918. The Eagle was installed in fifty other First World War aircraft types. The success and demand for the Eagle engine soon outpaced the Rolls-Royce production facilities. The USA was approached to build the engine under licence, but it never happened as the US Government backed its own Liberty engine, a 27-litre water-cooled 45-degree V-12 engine. Some contracts were signed to manufacture Eagle components, but the war ended, and Rolls-Royce was left with many agreements to be dissolved.

Engine	Rolls-Royce Eagle VIII
Horsepower (hp)	360
Number of pistons	12
Engine Capacity	20.33 litres
Bore	4.5 in (114.3 mm)
Stroke	6.5 in (165.1 mm)
Basic Dry Weight	847 lb (384.2 kg)

The Eagle engine can be seen at the Science Museum, London, and at Canada Aviation and Space Museum, Ottawa.

Rolls-Royce had entered a new era in its illustrious history, competing in the aero engine industry. After that first flight in a Handley Page 0/100 on 18 December 1915 at Hendon, London, by 1917 it had been installed in the Handley Page 0/400 heavy night bomber, the Airco DH4 day bomber and the F.2 and F.3 Felixstowe flying boats. A total of 4,681 Eagle engines were built, versions I–VIII, VIX in 1922 and XVI in 1925 (sixteen cylinder). What a success story for Rolls-Royce's entry into the aero engine world. The Eagle's success story did not stop at the end of hostilities. Eagle engines powered the Vickers Vimy of Alcock and Brown on the first successful non-stop crossing of the Atlantic Ocean on 14/15 June 1919. The Eagle engines had run for approximately sixteen hours. They also powered the first successful flight from England to Australia in a Vickers Vimy with total flying time of 135 hours. The Portuguese Navy successfully crossed the South Atlantic in a Fairey 111 floatplane with an Eagle VIII engine. Thus the Eagle began a series of 'birds of prey' Rolls-Royce piston engines.

Rolls-Royce Hawk

The Hawk was designed in 1915 and was derived from one bank of cylinders from the Rolls-Royce Eagle engine. It was a liquid-cooled, vertical inline, ungeared, six-cylinder engine. Initially developing 75hp, it was increased to 105hp by the end of the First World War. Rolls-Royce made the prototype and Brazil Straker continued to build the Hawk under licence to allow further work by the company on other projects. More than 200 engines were built, and it powered aircraft such as the Avro 504K, and SSZ coastal patrol airships. The Hawk was particularly noted for its reliability, confirmed by a 50 hours and 55 minute flight in August 1918. Routine RNAS patrols of up to 30 hours in the non-rigid airships were common.

Engine	Rolls-Royce Hawk I
Horsepower (hp)	75
Number of pistons	6
Engine Capacity	7.41 litres
Bore	4.0 in (101.6 mm)
Stroke	6.0 in (152.4 mm)
Basic Dry Weight	387 lb (175.5 kg)

The Hawk engine can be seen at the Shuttleworth Collection, England.

Rolls-Royce Falcon

The Falcon was designed in 1915 and it was a smaller version of the Rolls-Royce Eagle. It was a liquid-cooled V-12 design. The engine was designed by R. W. Harvey-Bailey and built under licence by Brazil Straker of Bristol, England. More than 2,100 were built and they were installed, among more than twenty-five other First World War aircraft types, in the Bristol F2B fighter (which in 2017 can be seen flying at the Shuttleworth Collection, Old Warden Aerodrome, Bedfordshire,

England). Not restricted to fighter aircraft, it also powered the Blackburn Kangaroo bomber. The engine featured an epicyclic propeller reduction gear with a clutch, which protected the reduction gear.

Engine	Rolls-Royce Falcon III
Horsepower (hp)	270
Number of pistons	12
Engine Capacity	14.2 litres
Bore	4.0 in (101.6 mm)
Stroke	5.75 in (146 mm)
Basic Dry Weight	715 lb (324 kg)

The Falcon engine can be seen at the Shuttleworth Collection, England.

Rolls-Royce Condor

The search for more horsepower continued not only with improved mechanical and scientific applications but the physical size of the engines. The Condor, first seen in 1918, too late to contribute to the war effort, followed the basic design of a 60-degree Vee-shaped twelve-cylinder liquid-cooled engine. The larger cylinders necessitated four valves per cylinder for efficient intake and exhaust gas flow. The Condor Series III went to a single spur reduction gear that required engineered strengthening to cope with transverse loads. By 1926 the oil, water and petrol pump were now housed in a single unit underneath the engine. Was this the first indication of streamlining considerations? To reduce weight, the hand starter mechanism was replaced by a Bristol gas starter. Six years later a Condor CI (C for civilian) engine was tested in a Hawker Horsley. This engine was unique in that it was a compression ignition, diesel engine that developed 500hp. Condor engines powered a variety of manufacturers' aircraft, such as Avro, Blackburn, de Havilland, Fairey, Handley Page, Hawker, Saunders-Roe, Short, Vickers and Westland.

Engine	Condor III
Horsepower (hp)	650
Number of pistons	12
Engine Capacity	35.03 litres
Bore	5.5 in (139.7 mm)
Stroke	7.5 in (190.5 mm)
Basic Dry Weight	1380 lb (628 kg)

Rolls-Royce Eagle XVI

This Eagle, no relation to the previous Eagle engine, was a 1925 experimental design of sixteen cylinders arranged in a revolutionary X design. When a matched supercharger and carburettor were borrowed from the Kestrel development programme, the engine ran well. However, it was not accepted by the aircraft industry as it was felt that the X arrangement would interfere with the pilot's forward view. The Eagle XVI never flew but the X design was revived for the Vulture and Exe engines.

Engine	Eagle XVI
Horsepower (hp)	500
Number of pistons	16
Engine Capacity	19.8 litres
Bore	4.5 in (114.3 mm)
Stroke	4.75 in (120.7 mm)
Basic Dry Weight	Unknown

Rolls-Royce Kestrel

By 1925 a radically new approach was developed using a single aluminium alloy monobloc for the two banks of six cylinders. Four valves per cylinder and a dry sump lubrication system completed the new engine. The F10, as it was first designated, was a direct drive extremely compact unit. It developed the same horsepower as the Condor, but it was approximately a third smaller in capacity and weight. By 1926 a geared version was available as the F12. The designations then changed to FXIA etc., as the Kestrel was quickly developed and improved upon by changing compression and gear ratios. In 1930 the 'F' type engine was named the Kestrel and initiated a Series, number I, II and III indicating gear reduction ratios. The suffixes A, B, were added to signify compression ratios and MS or S added to signify medium supercharger or full supercharger versions. The supercharger impeller was protected against sudden changes of crankshaft speed by the use of a special friction drive. By the end of the series' development, the Kestrel XVI was rated at 745hp at 3,000 rpm at 14,500 ft (4,420 m). It is interesting to note that as well as the Kestrel being installed in fifteen British aircraft, it was also installed in foreign aircraft for testing such as in the Messerschmitt Bf 109 and Junkers Ju 87 dive bomber. So the test results achieved would be used against the Allies four years later as the Second World War began.

Engine	Kestrel V
Horsepower (hp)	685
Number of pistons	12
Engine Capacity	21.24 litres
Bore	5.0 in (127 mm)
Stroke	5.5 in (140 mm)
Basic Dry Weight	957 lb (434 kg)

Rolls-Royce Buzzard

Originally known as the 'H' engine, the Buzzard was a similar design to the Kestrel but bigger and more powerful. The supercharger could be used to its full extent on the ground, so enhancing take-off performance. It was 50 per cent heavier and produced nearly 50 per cent more horsepower. It first ran in 1928.

Engine	Buzzard
Horsepower (hp)	955
Number of pistons	12
Engine Capacity	36.7 litres
Bore	6.0 in (152.4 mm)
Stroke	6.6 in (167.6 mm)
Basic Dry Weight	1540 lb

Rolls-Royce R engine

A development of the Buzzard engine, it was designed initially specifically for the Supermarine S6 seaplane competing in the 1929 Schneider Trophy Contest. It represents the very best of Rolls-Royce aero engine development in 1929 and of the British aero engine industry in general. The R engine for the same weight as the Buzzard nearly tripled the horsepower available, 265 per cent to be exact.

The pride of Britain was at stake and Rolls-Royce certainly rose to the challenge with Supermarine to put together a package to take part in, and win, the Schneider Trophy to show the world the advanced state of British aviation. The R engine had all the moving parts of the Buzzard, but it was fine-tuned to a 'horsepower means everything, longevity nothing' state! It did not even look like the Buzzard as streamlining had been taken to the extreme to fit into the streamlined engine compartment of the S6. Valve covers, cylinder blocks, engine covers were all adjusted for less drag. Auxiliaries were repositioned and, together with a lengthened reduction gear housing, further streamlined the engine frontal area. A newly developed supercharger took care of the increase in horsepower.

The 1929 R engine won the contest that year, but it was the 1931 improved version of the R engine that proved the competitiveness of the engine. It gave 2,300hp for a slight increase in weight, mainly achieved by increasing the engine speed, the supercharger gear ratio and the size of the air intake. Remembering that the R engine was built to deliver maximum horsepower and last just a short time, it was essential that it lasted the race, or else all was for nought. In April, the longevity was 20 minutes, in July it was 30 minutes and finally in August it was 60 minutes.

The engine was released as airworthy for the contest, which was won by Britain. On 9 September 1931, the R engine powered a S6B to a world record speed of 407.5mph (655.8kmh). The engine was also responsible for other records in the air, on the water and the land. It was the only engine to date to hold this distinction.

The engineering challenges were enormous and the stakes were high, not just the concerning national glory; there were ever-increasing intimations of a possible conflict in Europe. These were the foundation years of the Merlin engine, which was waiting on drawing boards to be brought to life. The ever-increasing loads of the R necessitated redesign of the crankshaft, connecting rods and crankcase. No part was exempt from improvement as maximum horsepower from minimum weight was sought. Imagine the stress and strain of a 9-ton (9,144 kg) load on the centre main bearing. That was what was occurring in the R engine. It took the strength of white metal on the main bearings to stand up to these stresses. Some of the metals that can make up a white metal alloy are antimony, tin, lead, cadmium, bismuth and zinc.

Other challenges included breaking valve springs, high oil consumption, the necessity of specific fuel blends and parts vibrating loose on magnetos. There was an enormous amount of research, design, development, resources and effort that went into preparing the R engine for the 1931 contest. This was, after all, as previously noted, the possibility of a third win in a row, which would give Britain the Schneider Trophy in perpetuity.

The Schneider Trophy now resides permanently in the Science Museum, London. It is a great monument to British aviation history and a forerunner of what would be accomplished during the hostilities just eight years away.

Engine	R (Sprint)
Horsepower (hp)	2,530
Number of pistons	12
Engine Capacity	36.7 litres
Bore	6.0 in (152.4 mm)
Stroke	6.6 in (167.6 mm)
Basic Dry Weight	1,630 lb (744 kg)

The R engine can be seen at the Science Museum, London; Royal Air Force Museum, London; Filching Manor Motor Museum, England.

Rolls-Royce Goshawk

Originally known as the PVG (Private Venture Goshawk?) the 1933 engine, based on the Kestrel, featured an evaporative or steam cooling system. This cooling system allowed the cooling liquid to boil rather than keep it below the boiling point. The theory applied was that by allowing the coolant to change form, from liquid to steam, it took more heat from the engine plus the added factor that by recirculating the coolant, less liquid was needed. The engine lost weight with less volume of coolant but lost efficiency in that a condenser was required to change the steam back to liquid. It was also quite bulky compared to a radiator and this caused extra aerodynamic drag. Problems with handling the coolant and the fact that the large wing-mounted condensers would be vulnerable to combat damage caused the Goshawk to be cancelled. Very few Goshawk engines were built, and they were only used in prototype testing as private ventures by the manufacturers. However, of historic importance, the Goshawk was installed in a Supermarine Type 224, which was the predecessor of the famous Spitfire.

Engine	Goshawk I
Horsepower (hp)	600
Number of pistons	12
Engine Capacity	21.25 litres
Bore	5.0 in (127 mm)
Stroke	5.5 in (140 mm)
Basic Dry Weight	975 lb (442 kg)

Rolls-Royce Merlin

Through these engines and the expertise of Royce and his team, the Merlin engine now appeared on the drawing board in 1933. It would turn out to be the greatest piston engine accomplishment of the Second World War by the Rolls-Royce Company. Undoubtedly hostilities contributed to its rapid design and continual improvement as the series developed to meet various applications. Who among aviation aficionados has not heard of the Merlin? The Merlin will be dealt with exclusively in the following chapters from its infancy in the 1930s to its continuing operation in 2018 and beyond. There will always be a Merlin running somewhere in the world as man cannot resist the challenge of having twelve cylinders and all its moving parts working in perfect harmony – and that is not even mentioning the sweet sound of the exhaust!

3

The Merlin
1933–34

When I was in my youth, I would read exciting accounts about the legendary Spitfire and its Merlin engine. At that time, I thought that using a wizard for an engine name added to the mystique. The legendary Arthurian wizard, called Merlin, with his supernatural powers and abilities, certainly stirred my imagination. I had read somewhere else that it was possible that Merlin was a sixth-century Scottish Druid. It was not until many years later, as I pursued my interest in aviation, that I realised that the engine was named after a bird of prey. In fact, it was the ninth engine to be named by Rolls-Royce after a raptor. The Eagle, Hawk and Falcon had preceded the name Merlin, which would become the most famous engine name of them all.

The merlin is a falcon found in the Northern Hemisphere; there are two species, the North American and the Eurasian. In North America, it is sometimes referred to as a pigeon hawk, not a suitable name for a piston engine. It is interesting that the American merlin males resemble the American kestrel, another bird of prey that has a Rolls-Royce engine named after it. It was the successful Rolls-Royce Kestrel engine that triggered the development of the larger Merlin engine. The merlin was very popular in medieval European falconry and, as per the engine, that merlin was heavier, had more muscle, and had greater speed and more endurance than the smaller kestrel.

The Merlin was born into the transitional period when the First World War biplanes and triplanes were coming to the end of their useful life and being replaced by the more suitable monoplane with stressed skin construction. There were exceptions, of course, mainly used in specialist roles, such as shipboard operation from the decks of small escort aircraft carriers. The Gloster Gladiator fighter and Fairey Swordfish torpedo bomber aircraft are typical of these last biplanes. The biplane construction provided the ruggedness that was able to withstand increasingly high g-forces until the development of modern engineering techniques, both material and structural, allowed the monoplane to cope with these same forces. The biplane had a limited life from the start because of the great aircraft aerodynamic trade-off, the basic physics of lift versus weight and thrust versus drag. The biplane offered lift at the great expense of weight and drag; the braces and wires that held the wings in place. Drag is the enemy of speed and is countered by the thrust of the engine and aerodynamic 'cleanliness'. Once the bracing was inside the wing and the wing area could be increased to give the same lift as the biplane, the monoplane started its own journey – which continues until today. Visibility in the omni-directional world of fighter aircraft was improved. The aero engines had increased in power to provide additional speed and manoeuvrability to the monoplane; now the lack of lift from the reduced wing surface area was solved.

Development of aviation after the First World War also faced the huge unanswered question of what the best type of engine for the application was – inline or radial, air- or liquid-cooled? The two types of engines each had their advantages and disadvantages, and it really became a balancing act for the aero engine manufacturers between their previous engine experience and the government's engine requirements. Power, weight, efficiency and reliability were the demands put on the internal combustion engine. The propeller was another matter. In researching this debate, I can only draw generalities and I leave it up to the reader to further research the topic, should he or she wish to delve deeper into the engineering science.

The inline engine presented a smaller streamlined frontal area and seemed more appropriate for high boost, high altitude and high-speed aircraft such as interceptors, fighters, and fast bombers. Its greatest vulnerability was damage to its cooling system by ground fire or an attacking fighter. This aspect of cooling efficiency is debatable. The rear cylinders on a radial engine are not cooled as well and the disturbed air adds to the total drag component. The liquid-cooled engine is more efficient because liquid is denser than air and can remove more heat. However, its nemesis can be the problem of getting uniform cooling despite the use of many thermostats. The air-cooled engine presents a larger frontal area, which can be detrimental to pilot visibility. However, it is more rugged and can develop more power at lower altitudes. It lends itself to use in low-altitude naval, ground-attack, and long-endurance patrol aircraft. Royce chose the inline liquid-cooled engine and bet the future of Rolls-Royce on it. Different engines were required for different applications.

What were these different applications in the 1930s as the Merlin was being developed? We have noted the transition from the biplane to the monoplane during this period but what of the military and civil aircraft types being developed by various manufacturers? What about the different engine manufacturers competing with each other to supply the power plants for the new aircraft? Everyone was trying 'to build a better mouse trap'. Who were these competing British aero engine companies and what were their engines called? Of course, other nations were creating engines for their domestic markets but with the hope that they could also sell their products abroad. Perhaps in Britain there was some protectionism present as well as national pride, which favoured the home product. This would give support to Rolls-Royce in the home market but how would it fare in the export market? However, it was unlikely, as the possibility of hostilities increased in the late 1930s, that Britain would be buying large numbers of foreign engines, certainly not from Germany, unlike prior to the First World War when Britain initially went to France seeking engines before the creation of the Rolls-Royce Eagle engine.

Some British aero engine companies were similar to Rolls-Royce, automobile companies that diversified into aero engines, while others were aircraft manufacturers that diversified into aero engines. Some ventures into the two areas were short-lived with the company having to decide on its manufacturing direction. The first thirty years of the twentieth century was a real trial-and-error time for the aviation industry. The experiences of the First World War, the desire of the wealthy to travel long distances quickly, and the threat of a second world war all drove the development of military and civilian aircraft. The open cockpit biplane was evolving into sleek monoplane designs and the quest for more engine power per weight and size. Of course, reliability was at the heart of the engine development and was a constant reminder of what could, and what could not, be done at the then current level of technology. Metal alloys, lubricants, carburettors, fuel injection, flame patterns, exhaust design and basic engine design were all developing, but the old adage applies – 'a chain is as good as its weakest link.' In other words, reliability.

What were some of these new British monoplane aircraft that demanded bigger and better engines? The Fairey Hendon was the first RAF all-metal monoplane night bomber, which flew in November 1930. The Hendon had Rolls-Royce Kestrel engines. The Vickers Wellesley and Handley Page Harrow had the Bristol Pegasus engine. The Bristol Blenheim and the Westland Lysander used the Bristol Mercury engine. The Avro Anson used the Armstrong Siddeley Cheetah engine. The Hawker Tornado used the Rolls-Royce Vulture engine. The Vickers Wellington and Saunders Roe flying boat used the Bristol Hercules engine. The Handley Page Hampden used the Bristol Pegasus engine. The Armstrong-Whitworth Whitley, Ensign and the Fairey Battle used the Rolls-Royce Merlin engine.

It is not surprising that the American Boeing 247, Lockheed Model 10 Electra, the Douglas DC2 and DC3 used the American Pratt & Whitney engine but it *is* surprising that the Vickers Warwick used the American engine. It was quite the mix and match of aircraft manufacturers and aero engine companies.

Famous names such as Armstrong Siddeley, Avro, Bristol, de Havilland, Fairey, Metropolitan-Vickers, Napier, Royal Aircraft Factory and Sunbeam were all major active competitors of Rolls-Royce in the British aero engine scene after the first Rolls-Royce aero engine, the Eagle, went into production. Other companies that were active were the Aircraft Disposal Company, the All British Engine Company, Alvis, Blackburn, Cirrus and Wolseley Motors. The growing industry was very competitive, and it would take exceptional talent and leadership to survive, but Rolls-Royce had both. What was the state of the company in the 1930s as the Merlin was coming off the drawing board and becoming a reality? The 1920s in Britain, tied to the gold standard, was in a period of depression, with deflation and general stagnation; this followed the post-war boom of 1919. Unemployment was high, greater than 10 per cent, and at the end of the decade the stock market crash of 1929 precipitated a global recession. One result of this recession for Rolls-Royce was the closing of the American Springfield plant in 1931. The company survived the 1920s and was looking forward to growth and prosperity in the 1930s and the Merlin would be a large part of that future.

Rolls-Royce management had a drastic shock in 1926 when Claude Johnson died and it had a difficult ten years until Ernest Hives took over the helm in 1936. The decade was not without its triumphs though. Rolls-Royce started the 1930s with the acquisition of Bentley Motors Limited in 1931, a formidable competitor in the luxury sport automobile industry, which would add to the company portfolio. In the same year, Britain won the Schneider Trophy in perpetuity mainly due to the superb Rolls-Royce R engine. The decade was marred by the catastrophic death of Rolls-Royce's second founding member, Henry Royce, in April 1933.

The company's growth, at this time, unfortunately, was due to the increased requirements of Britain's rearmament programme as Germany, Italy, Spain and Japan eventually withdrew from the League of Nations. The League of Nations was formed after the First World War to promote world peace and avoid future conflicts. After its failure, following the Second World War it was reformed as the United Nations.

In 1935, for the first time in its history, the Rolls-Royce aero engine division financially outperformed the auto division, and would continue to outperform it for the remainder of the decade. Regarding the previous discussion on the emerging monoplane aircraft, King George V reviewed thirty-seven RAF squadrons, comprising 356 aircraft, at RAF Mildenhall in 1935 and all the aircraft on display were biplanes; this was only four years prior to the start of the Second World War. That same year the Merlin, then called PV12, flew for the first time in a Hawker Hart at Hawkers, Martlesham Heath, and subsequently at the company test facility in Hucknall, Nottinghamshire. The first sixty engine hours were devoted mainly to adjustments of the cooling system. The 1930s saw the addition of Rolls-Royce shadow factories at Crewe, Cheshire and Glasgow, Scotland. Rolls-Royce was starting a modest expansion, which would soon grow exponentially as the demands of approaching war were met.

The Hawker Hurricane had its first flight in 1935 with the Spitfire following shortly after in 1936. Both these aircraft were powered by the Rolls-Royce Merlin engine and, as Britain's frontline fighters, would be built in large numbers and be instrumental in the defence of the island nation. 1937 saw partnership with Bristol in the new ROTOL Airscrews Limited. This year also saw the first order for the Supermarine Spitfire to be completed by spring 1939. There could be no doubt that the demand for engines would be great and Rolls-Royce had to some extent prepared for the increase in production. In Britain, the main and shadow factories would have to rise to the occasion, together with the additional American Packard Merlins built in the USA. The Merlin would go through many improvements and series to continue to produce more and more horsepower and not compromise reliability.

Notwithstanding the formation of the Rolls-Royce Company, the Merlin story began with the decision by Royce to, firstly, build his own design aero engine, and secondly, to make it a liquid-cooled engine based on his auto engine experience. There were twenty years separating the Eagle engine's first flight in 1915 and the first flight of the Merlin in 1935. Those years witnessed a myriad aero engineering innovations that were tried and accepted, discarded, or modified, prior to the Merlin becoming airborne. With the approaching hostilities, the emphasis during the 1930s was changing from the auto division to the aero engine division government contracts. The Merlin had a few variations along the way in the number and formation of cylinders and, importantly, the constant improvement of every part regarding design, size, weight, metallurgical content, strength, cost and ease of manufacturing; these just being a sample of the many engineering considerations.

The Rolls-Royce aero engine story, in reality, started with the auto engine and subsequently the two industries developed simultaneously from 1915. As we have seen, Royce had replaced his quadricycle with the French Decauville 10hp water cooled car. The Decauville started out as a 3hp air-cooled engine and progressed to the 8hp and 10hp water-cooled engines. It was the Decauville 10hp that Royce redesigned, and this was the first Royce water-cooled engine, the Royce 10, which first ran in 1903. The twin cylinder 1.8-litre 10hp engine started a series of ever more powerful auto engines beginning in 1904. The three cylinder, 3.0-litre, 15hp; the four cylinder, 4.1-litre, 20hp; and six cylinder, 6.0-litre, 30hp engines were all produced in 1905 and 1906. The exception to the series was the eight-cylinder V8 3.5-litre engine in 1905; few were made, and none survive.

The next engine, the 40/50hp, was a straight six, initially 7.0-litre, engine built from 1906 until 1926 for the Silver Ghost. Nearly 8,000 engines were built, and it established Rolls-Royce as a manufacturer in the big leagues. It was during 1915 that Rolls-Royce turned its attention to developing its own aero engine, the Eagle. By 1922 in the post-war era the company developed an inline six-cylinder 3.1-litre for the Twenty car. By 1925 a new engine was being manufactured, the 7.7-litre overhead valve engine for the Phantom car, which was built until 1931. A smaller, six-cylinder 3.7-litre 25hp engine was introduced and made the 20/25 car very successful. The auto engine developed simultaneously with the development of the Merlin was the six-cylinder 7.7-litre crossflow cylinder head for the Phantom II car built from 1929 to 1936. The design and development of all these auto engines would contribute to the design and development of the aero engines and vice versa.

Royce's first aero engine, the Eagle, was based on the engineering experience gained with the development of the 7 litre 40/50hp auto engine. Without this experience, it is doubtful whether the Eagle would have gone into production as quickly as it did. The liquid-cooled Eagle had two banks of six cylinders mounted in a 60-degree V formation, V12, a forerunner to the Merlin configuration. It needed the extra six cylinders to bring the government required horsepower to 200. By the end of hostilities, the Eagle engine had produced 360hp and this first successful Rolls-Royce aero engine had been installed in more than fifty different aircraft. The following aircraft were some of those powered by the Eagle – the de Havilland DH10, Hawker Horsley, Handley Page 0/400, Short Bomber, Supermarine Sea Eagle and Vickers Vimy. These same companies installed the Rolls-Royce Merlin in some of their aircraft twenty or more years later. It was certainly a long and profitable association. What a start for the Rolls-Royce Aero Division, which had already contributed, through its first engine, to the financial health of the company; the future looked promising.

The next engine, the Hawk, was derived from one bank of the Eagle engine. The smaller inline six-cylinder was installed in airships and the Avro 504, and thus began a long Avro association with Rolls-Royce culminating in the famous Merlin engine in the Avro Lancaster. The Falcon was a smaller version of the Eagle and it equipped the Bristol F2B fighter among others. The Merlin engine was installed in the Bristol Beaufighter. The Condor engine followed; it was massive, like its namesake bird. Westland joined the Rolls-Royce family with the Condor and subsequently later joined the Merlin engine club. The Eagle and Condor engines as per the 1920s had separate cooling jackets for each cylinder, which would shortly change. Sir Richard Fairey, founder of the Fairey Aircraft Company in 1924, felt that there were no British engines suitable for his aircraft and went to the USA looking for an engine. He returned with an agreement to build the Curtiss D12 under licence. This was an aluminium alloy monobloc cylinder engine with wet cylinder liners. The turning point in the foundations of the Merlin engine was in 1925 when Rolls-Royce was asked by the Air Ministry to examine the Curtiss engine and manufacture one similar to it, which resulted in the Eagle XVI and then mid-year in the radically different F-X, or F, engine being developed. The 'turning point' is in dispute as Royce had started his design the year before seeing the D12, but perhaps it gave him some ideas for his own engine. The single aluminium alloy monobloc, with four valves per cylinder, heralded a new direction in engine design. It had a single camshaft that operated the valves by rocker arms.

Testing began in 1926 and three faults became immediately apparent. The dry liners were replaced by wet liners similar to the Curtiss D12, a reduction gear was fitted, and the supercharger was temporarily abandoned, the engine run normally aspirated. 1927 was a year of continuing testing and by summer, the engine had completed an endurance 100-hour test at 2,100 rpm producing 450hp. Rolls-Royce was advised to hire J. Ellor, a supercharger expert who became the Rolls-Royce Chief Experimental Engineer. The same year, Lieutenant Colonel Fell took over the company's Flight Test Centre at Hucknall. Rolls-Royce had accumulated significant numbers of experienced personnel over the previous years and the company had entered an age of advancement within the aero engine industry. Installed in the Hawker Hart and a family of Hawker variants, the engine, now known as the Kestrel, was another success story for Rolls-Royce. It was 20 per cent smaller than the Merlin and was instrumental in the Merlin's design to provide relatively more power from a larger engine.

The Rolls-Royce Buzzard was approximately 40 per cent larger than the Merlin and it was developed into the R engine to compete for the Schneider Trophy. The R engine won both the 1929 and 1931 trophy races, in the Supermarine S6 and S6B respectively. The R engine had achieved unthought of values of performance, 2,300hp in 1931 with piston speeds and cylinder pressures never attained before. These achievements prompted design studies known in the Works as the PV12. The steam-cooled Goshawk was abandoned and the PV12 – 12 for twelve cylinders – was originally considered in April 1932 to be built inverted for pilot sight lines, but the airframe manufacturers vetoed the idea. The project proceeded with the Air Ministry kept informed of its progress, bearing in mind that Rolls-Royce was funding this whole development and taking a chance on the outcome that it would be a winner, a risk that paif off, as history would record. At this time, Rolls-Royce was battling other manufacturers for market share in the auto and aero industries. The last minute outbidding in 1931 of the Napier aero engine company for the Bentley car company is an example.

One of Royce's last decisions before he died was to authorise the construction of an engine larger than the Kestrel but smaller than the Buzzard, incorporating all the lessons of the R engine. Out of all these previous great engines, the PV12 (Merlin) appeared on the Rolls-Royce drawing boards and first ran on 15 October 1933, six months after Royce's death and a mere six years before the start of the Second World War when it would excel as the engine of choice for Allied fighter and bomber aircraft. The PV12 was a result of development of past engines, brilliant thinking of many minds, and the testing of components and the full-scale engine. It was a logical progression from the Kestrel and R engines. The first two PV12 engines were numbered one and three. The Rolls-Royce convention was to number odd serial numbers for 'left hand tractor' engines that rotated anti-clockwise when viewed from the rear and even numbers for 'right hand tractor engines' that rotated clockwise when viewed from the rear.

At the same time as the birth of the Merlin, the global community was struggling with how to avoid another world war. The Women's International League for Peace and Freedom (WILPF) grew out of the Congress of Women, a 1915 gathering of women to protest against the First World War. The women, who had come from all parts of Europe to work for peace, crafted twenty resolutions in an attempt to bring warring nations to the peace table and end the First World War. In 1932, working with its Nobel Peace Prize-winning president, Jane Addams, WILPF members collected six million signatures for the World Disarmament Petition and delivered them to the World Disarmament Conference in Geneva. The word 'disarmament' in the title was not a realistic goal but the thought was that all parties would settle for arms limitation instead. For Rolls-Royce this meant that the conference might limit the size of offensive weapons, such as bombers, which would limit the requirement for larger, more powerful engines than the Rolls-Royce engines presently available. The targets for bombers were industrial areas and cities.

There could not be a limit on military aircraft alone, as civilian aircraft could be converted to military use quite quickly. The suggestion was to give up military aviation, including naval aviation, entirely, and to eliminate aircraft carriers. Japan had invaded Manchuria in the early 1930s and was so confident that no country or countries would oppose it that it withdrew from the League of Nations in 1933 because the League would not recognise the puppet government it had established there. Six months later, Adolf Hitler, Chancellor of Germany, withdrew from the League of Nations and ordered that the German secret rearmament process be stepped up in spite of the Treaty of Versailles restrictions. The conference continued until 1934 with no agreements and it faded into history, unable to stop the Second World War. There would be no restrictions for Rolls-Royce in their engine development programme; the testing continued.

The PV12 had an ignominious start, with the 100-hour test showing disappointing power output results. Design changes were required to make the engine viable, remembering this was a private venture, meaning no performance, no revenue. Testing also revealed that the Merlin reduction gear had to be redesigned and that the block casting needed strengthening. By July 1934, the PV12 had passed its first test and, at 1,030hp, achieved a rating of 790hp. The innovative design approach continued and a new 'ramp' head cylinder was installed, the aim being to improve flame travel by incorporating two flat ramps of unequal size and inclination in the combustion chamber to create more turbulence of the fuel-air mixture.

This engine was designated as the Merlin B engine and the 'ramp' head was incorporated in the Merlin B to F model engines. Built in the autumn of 1934, the Merlin B engine delivered 950hp on test in February 1935. The ramp head never achieved its expectations and would eventually be abandoned; the problems included local detonations, cracks in the cylinder head and exhaust valve fractures. Continuing the development and design changes to improve the engine, Rolls-Royce decided to change the present monobloc casting construction and to cast the crankcase and block as separate units. These engines were called the Merlin C and the aim was to simplify the manufacturing process and improve the quality of the castings. This engine was used in the Hawker Hurricane prototype.

These B to F engines were experimental and did not go in to production until the Merlin F engine, the Merlin I. The G engine replaced the 'ramp head' with the 'parallel pattern', valves parallel to the cylinder head and this was known as the Merlin II. By the autumn of 1937 the Merlin II had passed its acceptance test and at 1,030hp the production line had increased the output of the engine that powered the Hurricane, Spitfire, Defiant and Battle.

Rolls-Royce was not the only company vying for Air Ministry specification requirement orders. The Merlin was born in a time of increasing competition to meet the new requirements of the RAF in its transition to acquiring new, more powerful fighters and bombers. Other aero engine companies were designing, testing, discarding, producing, and developing aero engines of all shapes, technical innovations and sizes. Bristol and Napier, up until now, perhaps had a slight edge on Rolls-Royce, but that would all change with the Merlin. The old question of air-cooled or liquid-cooled still puzzled the engineers as they tried different engine configurations, all in the quest to get the highest horsepower to capacity and weight ratio.

During the early 1930s Armstrong-Siddeley, Bristol, Fairey, and Napier were all competition for the Merlin with their own individual approach to engine development. The engines are indicative of the development years before the Second World War, with the stated horsepower increasing as the engines were improved. The Fairey Prince and Monarch never went in to production.

Armstrong Siddeley	1935 Cheetah	7 cylinder 13.65 litres		
		637lb/289kg		307hp
Armstrong Siddeley	1929 Panther	14 cylinder 27.3 litres		
		1047lb/475kg		638hp
Armstrong Siddeley	1932 Tiger	14 cylinder 32.7 litres		
		1287lb/584kg		850hp
Bristol Perseus	1932	9 cylinder 24.9 litres		
		1025lb/465kg		905hp
Fairey Prince	1939	16 cylinder 34.05 litres		
		2180lb/989kg		1540hp
Fairey Monarch	1939	24 cylinder 51.08 litres		
		2180lb/989kg		2240hp
Napier Rapier	1929	16 cylinder 8.833 litres		
		720lb/327kg		1340hp
Napier Sabre	1938	24 cylinder 36.65 litres		
		2360lb/1070kg		2200hp

On the Axis side, aero engine development was keeping pace.

Daimler-Benz DB601	1935	12 cylinder 33.93 litres	
		1300lb/590kg	1175hp
Junkers Jumo 213	1940	12 cylinder 35 litres	
		2072lb/940kg	1726hp

By 1935 the Merlin was rapidly approaching its first airborne test. By then it was fairly obvious that another conflict was imminent. It would take until autumn before the Merlin was ready to go flying; there had been a few setbacks during the year. PV12 Number 1 engine was installed in a Hawker Hart and went airborne for the first time on 12 April 1935 at the Hawker facility at Martlesham Heath. This particular aircraft had an evaporative cooling system similar to the one tried with the Goshawk engine, which had been a failure due to coolant leaks. The Hart was converted to a 100 per cent glycol system and continued the engine testing at the Rolls-Royce Hucknall facility. This Number 1 engine's power output had already been exceeded by the Merlin C engine by 255hp.

Pressure was mounting both on the government to get prepared by examining their future requirements and on the manufacturing companies to be ready for the forthcoming orders. Up until 1935 the procurement procedure was very inefficient, which resulted in an unacceptable length of time between the requirements being issued and actual production beginning. The initial solution was simplicity itself: two prototypes to cut the testing time in half, at least that was the intention. A major decision, once again political circumstances in Europe dictating, was to allow new orders for aircraft before the prototype had been tested. This had a major influence on the production of the Armstrong Whitworth Whitley with the later Model V having Merlin engines.

The Special-Order system placed the design and production of the requirement with the company that the Air Ministry felt was the most capable. The companies in turn were very aware of what the Air Ministry required. This procedure was known as 'Private Venture' and certainly put the risk on the manufacturer. The Air Ministry, if interested, would purchase the prototype and pay for its development. If the Ministry was not interested, the company had gambled and lost with their idea, a costly venture. The PV12 was the gamble that became the Merlin.

The previous chapters and the beginning of this chapter gave the background and history of Rolls-Royce and its aero engines leading up to the construction of the Merlin engine. The rest of this chapter will give a general overall description of the engine and its components but please bear in mind that the Merlin was constantly evolving over the years; it is not the intent of this book to describe in detail all aspects and modifications of the Merlin engine. The components described are generally associated with the Merlin 60/70 series of engines but other components from other series may be included. The Merlin was built not only by Rolls-Royce but also under licence by the Packard Motor Car Company in the USA.

THE MERLIN BASIC CONFIGURATION

The Rolls-Royce Merlin is a glycol-based liquid-cooled V12, piston aero engine, of 27.0 litres (1,650 cu. in.) capacity with a five point four-inch stroke and a six-inch bore. The Packard version of the Merlin was called the V-1650. It should be mentioned at the outset that the engine is a vast network of shafts and gears, of all types, shapes, descriptions and sizes, to operate the engine and drive the components. Intricate hardly describes the mechanical setup.

ENGINE

CYLINDERS: The two-cylinder assemblies, block 'A' the starboard and block 'B' the port, comprise six cylinders each, the upper camshaft drive unit and the camshaft and rocker mechanism operate the valves in the cylinder head. Each block has a separate light alloy skirt bolted to a head to form the cylinder block. The head is the roof of the six combustion chambers. There are six detachable wet steel liners in the skirt, which is also part of the coolant jacket.

Seven brackets are mounted on the top of both cylinder blocks, with each bracket having a bearing to support the camshaft. The camshaft operates the valves by a separate cam and rocker for each. The inlet and exhaust valve rockers are each mounted on their own respective spindle. A cam bears on each rocker and the rocker operates a valve through an adjustable tappet. The two spindles extend rearwards to carry auxiliary driving wheels for an air compressor mounted on the 'B' block and an air compressor or hydraulic pump mounted on block 'A'. A third spindle, the upper camshaft driving spindle, is driven through a separate coupling from the wheelcase. The cylinder blocks are covered by a cast aluminium cover, which includes the rpm indicator drive on the 'B' side.

CYLINDER BLOCK: The skirt, lower half, has the liners with a coolant space provided between each cylinder and between the combustion chamber and the top of the block at the head. Four gas passages lead outwards from the combustion chamber to the side of the head. The inlet passages connect with a common elongated port in the inner face to which the induction manifold is attached. The two exhaust valve passages are combined in an individual port on the outer face of the head. The six exhaust ports have a branch pipe exhaust manifold. The valve seats are made of high-silicon-chromium steel screwed into the mouth of each respective passage, inlet or exhaust, where it enters the combustion chamber. A spark plug aluminium bronze adaptor at right angles to the cylinder is locked by a left-hand screwed ring in a boss in the cylinder head.

The top of the cylinder liner fits into a recess on the combustion chamber. Aluminium connections and rubber sealing rings form cooling integrity of the liners, skirt and head. Twenty-four studs secure the cylinder head to the skirt. A further fourteen studs secure the cylinder block to the crankcase. Oil is drained from the valve mechanism to the crankcase by guard tubes and the upper camshaft drive tube.

CYLINDER LINER: A gas-tight joint is maintained to the combustion chamber with a shouldered and spigoted top end and a sealing ring at its lower end to form a joint with the crankcase.

VALVES: The inlet and exhaust valves are not interchangeable. They are both trumpet type and have stellite-ended stems. The inlet valve stems are solid with the seating facings surfaced with Brightray while the exhaust stems are hollow sodium-cooled. A circlip prevents the valve falling into the cylinder in the event of a broken spring mechanism. The springs are a pair of concentric coil springs that return the valve to its seat. The valve guides are also not interchangeable. The inlet ones are cast-iron while the exhaust ones are phosphor-bronze.

CAMSHAFT: A single central camshaft mounted in seven pedestal brackets operates the inlet and exhaust valves by rockers fitted with adjustable tappets. The rockers pivot on side shafts. The similar camshafts, block 'A' and 'B', are driven from the wheelcase by inclined shafts ending in a bevel mechanism at the rear ends of the camshaft. This bevel mechanism in turn drives auxiliary driving spindles mounted on the rear of the block. These spindles and gearing drive air compressors or hydraulic pumps. Lubrication of the camshaft, cams and bevel gear is by the low-pressure oil system through the hollow camshaft, camshaft pedestals and rocker spindles.

The cams are formed in groups of four for the two inlet and two exhaust valves per cylinder. The inlet valves are the inner positions. Each camshaft is of nickel steel and has radial holes to distribute the low-pressure oil. The rear of the 'B' side camshaft has a gear wheel to drive the rpm indicator and the 'A' side is plugged. The upper crankshaft drive is driven from the wheelcase by a guarded coupling shaft, which also allows oil to return to the crankcase.

ROCKER: The rocker fulcrum shafts are located on the opposite side of the block to the valve it is operating. The hollow shaft is plugged at the front and the rear of the shaft projects beyond the bracket to the auxiliary driven gears. The rockers are of two patterns, left and right hand. The end of the rocker arm has a tappet screw to adjust the clearance between the cam bases and pads. Oil is supplied from the hollow pivot shaft to the rocker mechanism.

EXHAUST MANIFOLDS: The 'A' and 'B' exhaust manifolds are similar in design but are left- and right-handed. They are sheet steel pressings, which are then welded to form each component. Each manifold section is interconnected with tubular push fit extensions. The pipes formed are lengthened from front to rear. The outlet from each section is a rearward facing nozzle, which provides an injector effect. There are three types of manifolds fitted to the Merlin depending on aircraft requirements, the fish tail three-outlet orifice type with a gun heating tube passing through them, the crescent and circular shaped, rear outlet only, orifice type and the individual ejector mounted on each exhaust port. On the bomber Merlin marks there were flame-damping exhausts fitted.

PISTONS: The pistons are attached to the connecting-rods by fully floating gudgeon pins. The connecting-rods are of two types, the plain fitted to the 'A' side and the forked fitted to the 'B' side. The connecting rods are assembled in six pairs. The big-end bearing or split bearing block for the forked rod is in two parts, held to the rod by four bolts. Steel shells lined with lead-bronze are used for both types of rods. The pistons, machined from light-alloy forgings, have three compression rings above the gudgeon pin and a grooved radial holed scraper ring below. The rings are free to rotate, no stops are fitted.

CRANKSHAFT: The crankshaft is of the six-throw balanced type and provides drive for the reduction gear and wheelcase components. The throws are paired one and six, two and five, three and four in three planes, 120 degrees apart in the direction of crankshaft rotation. Eight of the crankwebs are extended to carry integral balance weights. The crankpins are numbered sequentially from front to back, one to six. The journals and crankpins are hollow providing an oil reservoir, which provides lubrication to the connecting rod bearing surfaces. There is a flange and coupling at the forward end, which drives the reduction gear and has valve and ignition timing marks on it. Timing operations are carried out solely by the B1 and A6 cylinders. At the rear end the crankshaft has a coupling to drive the spring-drive unit.

CRANKCASE: The two-part crankcase incorporates the rear half of the reduction gear housing at its front end and has a facing at the rear end to which the wheelcase is attached. The upper portion is an aluminium-alloy casting of pentagonal cross-section and has two top faces for the inclined cylinder heads, 60 degrees to each other, giving rise to the V nomenclature, as in V12. Each face is drilled with six spigot sockets to take the cylinder liners. The four engine mounting feet protrude from each corner of this casting. There are webs at the ends and five internally, which divide the crankcase into six compartments for two cylinders each compartment.

The lower crankcase portion forms an oil sump and is drained by scavenge pumps in the depressed floor at the rear of the casting. Two detachable filters are located here with the oil pressure pump being located underneath the rear scavenge pump. The relief valves unit and the electric generator, right side, and intercooler pump drive housing, left side, are mounted on crankcase facings. Two brass plates on the port side indicate the engine data plate and the cylinder firing order.

UPPER CRANKCASE: The reduction gear facing is faced concentrically with the propeller shaft axis. There are two bearing housings in the extended forward end of the crankcase. The lower one is concentric with the crankshaft axis and the upper one with the propeller shaft. Both these bearings have outer roller races, which are joined by the reduction gear driving pinion. Immediately behind the propeller shaft bearing housing is an oil gland, which connects propeller pitch operating oil to the propeller shaft. Depending on type of propeller fitted, the oil originates with the oil relief valves unit or the constant-speed unit. The crankshaft bearings are in two flanged steel shell halves lined internally with lead bronze fitted in the housings formed by the crankcase and bearing caps. The lower halves have two holes fed by oil from the bearing cap. The main pressure oil is distributed by a light-alloy pipe to the crankcase webs and bearing caps.

LOWER CRANKCASE: A trough-shaped aluminium-alloy casting with a shallow sump at the front and a deep sump at the rear end which is faced for the wheelcase. A baffle forward of the oil pump and filters assists with the drainage of oil thrown off by the connecting rods.

HYDRAULIC PUMP DRIVE [CRANKCASE]: It is attached to the bottom of the lower crankcase and the drive, vertical or combination vertical and horizontal, is attached in front of the oil pressure pump.

REDUCTION GEAR [NON-CABIN SUPERCHARGER TYPE]: It is of a single spur layshaft type and the casing is partly in the crankcase itself. The driving pinion is concentric with, and is driven by, a short coupling shaft from the crankshaft. The extended (to fit the hub) nickel steel universal propeller shaft is above and parallel to the coupling shaft. Both shafts run in roller bearings.

PROPELLER PITCH GOVERNOR & VACUUM PUMP DRIVE: The dual-drive unit fitted to the lower part of the reduction gear cover has a constant-speed propeller governor and a vacuum pump for navigation instruments. Lubrication is achieved by splash oil and inlet supply oil to the governor unit.

WHEELCASE ASSEMBLY: It is bolted to the rear end of the crankcase and drives and carries many of the important components of the engine. Notwithstanding the power generation components, the pistons and crankshaft, it really is the heart of the engine keeping the whole machine running.

THE WHEELCASE CARRIES the magnetos, coolant pump, generator drive, wheelcase vent, supercharger casing, the hand and electric turning gear and the fuel pump unit.

THE WHEELCASE HOUSES the spring-drive unit and certain shafts driving the magnetos, camshafts, electric 12-volt generator, fuel, oil and coolant pumps. The spring-drive unit is a torsionally flexible shaft through which various shafts and wheels in the wheelcase are driven. The rear portion houses part of the impeller drive.

SPRING-DRIVE UNIT: The unit transmits the drive from the crankshaft to the supercharger and auxiliary drives. It comprises a solid inner shaft and a hollow outer shaft. A bevel wheel splined to the centre of the outer shaft drives the upper and lower vertical drives. The lower vertical drive is connected to the fuel pump spindle, the oil pump idler gear, and the coolant pump rotor spindle. The upper vertical drive shaft is connected to both magnetos via a transverse shaft and the lower camshaft drive.

COOLANT PUMP: It is a centrifugal-type coolant pump that serves both cylinder blocks from two diametrically opposed separate outlets. It is driven by the lower vertical shaft. The impeller shaft sits in a self-lubricating Morganite thrust pad bush.

INTERCOOLER PUMP: The pump is mounted on the rear of the generator drive on the port side of the crankcase. It circulates coolant from the header tank to the radiator. It is driven by spur gears from the generator drive shaft. The pump comprises three main casings.

FUEL PUMP ROLLS-ROYCE GEAR TYPE: The pump unit is mounted on the port side of the wheelcase. It consists of two pumps working in parallel. Each pump is driven separately by a shear coupling shaft from the lower vertical shaft and operates independently of the other. Each pump is capable of supplying maximum engine capacity demands on its own. Oil is supplied under pressure for lubrication purposes. An interchangeable fuel pump is the vane type Pesco F8Mk1.

ELECTRIC GENERATOR AND INTERCOOLER PUMP DRIVE: They are driven both by the left-hand supercharger layshaft through a gear-train.

HAND & ELECTRIC TURNING GEARS: The engine may be turned, started, by hand or by an electric motor located within the starboard side of the wheelcase. A turning handle is inserted for a manual start. There is a clutch mechanism to disengage both starters.

SUPERCHARGER ASSEMBLY: The supercharger on the Merlin 66, 70, 71, 76, 77 & 85 Series of engines is a two-speed, two-stage, liquid-cooled tandem two-rotor centrifugal type. It is driven from the crankshaft through a two-speed gear at different ratios depending on the engine. For example, the Merlin 66 engine has a ratio of 5.79 & 7.06 to 1. There are separate spindles and centrifugally loaded clutches for the low and high gear drives. The gears are selected by selection of the oil pressure-operated servo-cylinder. The boost control unit and carburettor assembly are mounted on the double-entry air intake elbow. The two intake ducts merge over the central area of the first stage impeller. The supercharger mixture is then led through the intercooler to the induction trunk. The accessory gearbox is bolted to the port side of the supercharger front casing.

GEAR CASING ASSEMBLY: The casing contains the three driving clutch gear assemblies of the supercharger. There is a constant stream of scavenged oil to all the gears, pinions, layshafts and the driving wheel. A camshaft actuates the low & hi speed gear selector forks. The camshaft is activated by an oil-controlled servo cylinder, which in turn is activated by a change-over valve and this, in turn, is actuated by a manual control from the cockpit or by an automatic altitude control. The upper clutch driving wheel assembly provides the low gear ratio and the two lower assemblies provide the high gear ratio of the supercharger drive. The low and high gear are practically identical except that the high clutch gear, instead of being a separate clutch gear similar to the low gear, is integral with the periphery of the clutch casing.

INTERCOOLER ASSEMBLY: The two-stage Merlin engines have an intercooler fitted to cool the high temperature delivery mixture generated by the two-stage supercharger. The coolant is circulated by a centrifugal type pump. The assembly is located at the top of the wheelcase and connects the supercharger outlet and the central induction manifold. The supercharger mixture passes from front to rear in the intercooler while the coolant flows horizontally. The Merlin 66 has a ribbed plate instead of the cylindrical header tank and the coolant is carried in a tank over the reduction gear casing.

PIPING SYSTEM: There is an extensive array of external pipes on the engine. The pipes carry oil, air, coolant, fuel, fire-extinguishing and de-icing solutions.

CARBURATION
The operating pressure of the Bendix Stromberg carburettor is controlled by the weight of an air pressure-temperature sensing device, passing through it to control the fuel pressure difference on either side of a fixed jet. The fuel is then delivered to the supercharger intake eye via a discharge nozzle. When the induction pipe pressure exceeds the maximum economical cruising value, the fuel specific consumption is increased. Fuel flow during the idling range is mechanically controlled. The fuel is supplied to the carburettor by fuel pumps drawing fuel from the fuel tank and passing through the de-aerator and filter. A relief valve bypasses fuel at 15psi. There are two fuel pressure systems, the primary, unmetered, and secondary, metered. It is beyond the scope of this book to explain the intricate workings of the carburettor but suffice to say it includes components such as regulator fill valve, cut-off valve, jets, fuel enrichment valve, slow running system, discharge nozzle and diffuser, accelerator pump, and volute drain unit.

AUTOMATIC BOOST CONTROL: This control automatically maintains the boost pressure set by the cockpit throttle lever under varying intake pressure and speed conditions. The throttle can be set and left by the pilot at the required boost setting. It works on the basic principle of two opposing springs maintaining balance between them. Simultaneously the throttle adjusts the magneto settings.

INDUCTION: The supercharger mixture passes through the intercooler to the central induction trunk which is in the V of the cylinder blocks at the top of the crankcase. The trunk has six faced outlets connecting with the three faced inlets in each induction manifold. The induction system incorporates the priming and slow running system.

LUBRICATION

There are four main oil circuits in the engine, the main pressure, the low pressure, the front scavenge pump and the rear scavenge pump circuit. The main and low-pressure circuits are served by a single pump and relief valves. The scavenge circuits are each served by its own scavenge pump, giving an engine total of three oil pumps.

OIL PUMPS: They are gear type and have a pair of scavenge pumps mounted above the pressure pump separated by the floor of the crankcase rear sump. All the pumps are driven at an approximately 25 per cent reduced rate by a vertical shaft from the wheelcase.

SCAVENGE PUMPS: The deliveries of the two scavenge pumps are in parallel. The scavenge pump consists of a pair of gears in its casing and both casings being integral and secured to the inside floor of the crankcase.

SCAVENGE OIL FILTERS: Each scavenge oil filter has a separate bell-shaped aluminium-alloy casing enshrouding the filter gauze element for each scavenge pump.

PRESSURE PUMP: The pressure pump is secured to the outside of the rear crankcase sump.

RELIEF VALVES UNIT: The unit is mounted on the right-hand side of the crankcase. It comprises two relief valves and internal passages, compartments, external unions and connecting ports to route the oil to its various functions. These include low pressure to reduction gears, supply pipe for propeller control, oil thermometer union and low-pressure oil to wheelcase and supercharger.

MAIN PRESSURE CIRCUIT: It is served directly by the pressure pump, which gets its oil from the tank via a filter. The pump outlet total delivery volume, at approximately 80 pounds per square inch, goes to two locations, the relief valve for crankshaft, big-end bearings and main bearing lubrication and, for engines equipped with the constant-speed governor, the governor control unit on the reduction gear cover.

LOW PRESSURE CIRCUIT: The oil at 6 pounds per square inch is used for reduction gear oil jets and camshaft drives, supercharger casing, generator drive casing and impeller gear teeth.

FRONT SCAVENGE PUMP CIRCUIT: Served by the port pump, the drained oil from the reduction gear, camshaft and rocker mechanism, and crankshaft, collects in the forward sump and by way of a pipe, returns to the scavenge pump via a filter. The oil is then returned to the holding tank.

REAR SCAVENGE PUMP CIRCUIT: Served by the starboard scavenge pump, the drained oil from the wheelcase and supercharger components, crankshaft, camshaft and rocker mechanism, collect in the rear sump and returns to the scavenge pump via an oil filter and thence to the holding tank.

IGNITION

Two Rotax or BTH magnetos provide power for the twenty-four spark plugs. The magnetos are mounted on the port and starboard side of the wheelcase. The port or 'B' side magneto serves the spark plugs on the exhaust valve side of both blocks and the starboard or 'A' side magneto the inlet valve side of both blocks. The magnetos are driven by a drive shaft in the wheelcase. The timing is advanced as the throttles are advanced, retarded as they are closed, by means of a system of rods, levers and shafts connected to the contact breaker cover.

IGNITION CABLES: Each screened magneto cover has twelve screened metal braided high tension spark plug cables connected to its sockets. The cables are routed via three conduits, one in the centre between the blocks and two outside each block. Special spark plugs have hollow threaded extensions on them to facilitate quickly detachable, but locked-on, cable adapters.

The Merlin engine (there are far too many variants to be specific) is operated manually by a throttle and propeller speed control, and either manually or automatically for radiator shutters, mixture, boost and supercharger control. The Merlin had been built and flight tested and was now available for the manufacturers to examine and consider for their new sleek monoplane designs. Development and design changes would continue as the engine gained running time. The Air Ministry was aware of the engine and what Rolls-Royce was doing. What aircraft manufacturer would decide on the engine to power their new aircraft and would it be a fighter or bomber type? Did the engine have the basic design characteristics to enable it to be improved and produce more power? Did it have the ability to produce that power at high altitudes? One manufacturer thought so; Hawker aircraft, with Sydney Camm leading the design team.

PART 2

THE MERLIN'S FINEST HOUR
1935–1940
The Fighter Defence of Britain

4

The Hawker Hurricane & Supermarine Spitfire 1935–39

In the previous chapters the Merlin and its manufacturer, Rolls-Royce, its pedigree, the engines that contributed to its development, and the actual engine and its components have been explored. Now we will look at the end users of this great engine and what they accomplished with it to make it 'The Engine that Won the Second World War'. The engine was young at the beginning of hostilities, six years old, and so were the crews who used it as an instrument of war. The airline industry was also young; it flew the Atlantic Ocean during and after the war with Merlin-powered aircraft until the jet age took over. Would the Merlin reach sufficient maturity to contribute to the war effort? Could a sufficient number of engines be produced to keep up with airframe production?

My impression of the Merlin is it is like electricity, you turn the switch on and, without thinking about it, the light comes on. Similarly, with the Merlin, you start the engine and, without thinking, you carry out your task in whatever vehicle you are in, aircraft, ship, or tank, and expect the engine to perform to a certain standard. This did not just happen; a lot of people were involved to get the engine to this level and it was also expected that Rolls-Royce would continue to improve the reliability and power of the engine.

It was with this Merlin engine that men such as Bader, Malan and Lacey contributed to the Allied victory twelve years after the Merlin's birth. Group Captain Sir Douglas Bader CBE, DSO & Bar, FC & Bar, FRAeS, DL, was a fighter ace, more than five confirmed enemy kills, and he had flown both the Hurricane and Spitfire. Group Captain Adolph 'Sailor' Malan, DSO & Bar, DFC & Bar, was a South African fighter ace and one of his ten rules for air fighting was 'Go in quickly – Punch hard – Get out!'

James 'Ginger' Lacey, DFM & Bar, was one of the top scoring aces of the RAF. The *London Gazette* wrote 'Sergeant Lacey has shown consistent efficiency and courage. He has led his section on many occasions and his splendid qualities as a fighter pilot have enabled him to destroy at least 19 enemy aircraft.'

The Merlin is an inanimate though it does have something akin to its own lifeblood in air, petrol, oil and coolant. It was the personal toil of many people who guided it on its journey to success. The Merlin would reach its zenith by the effort of human brain and muscle power. Following the concept, design, engineering, testing and production of the Merlin came the administration, including the company, the military and the government, the human input to make the whole concept work. We must not forget the financial decisions that smoothed the way for the Merlin to come to fruition.

With any inter-neighbour dispute, the first order of business is to secure your own home and in Britain's case it was to secure the skies above the country. This is purely a defensive move and the political leaders were well aware of what was required in 1940. It would take a fighter or fighters to keep the skies clear and they would have to have the firepower to down bombers and the performance to out-manoeuvre the escorting German fighters. The German forces had swept into the Low Countries and Holland on mainland Europe and there was no doubt that the British Isles were a target to enable the securing of the whole of Europe for Germany's expansion. Winston Churchill had recently taken over as Prime Minister, and this is an excerpt from a speech that he made to the British Parliament on the 13 May 1940:

What General Weygand called the Battle of France is over. I expect that the Battle of Britain is about to begin. Upon this battle depends the survival of Christian civilization. Upon it depends our own British life, and the long continuity of our institutions and our Empire. The whole fury and might of the enemy must very soon be turned on us. Hitler knows that he will have to break us in this Island or lose the war. If we can stand up to him, all Europe may be free and the life of the world may move forward into broad, sunlit uplands. But if we fail, then the whole world, including the United States, including all that we have known and cared for, will sink into the abyss of a new Dark Age made more sinister, and perhaps more protracted, by the lights of perverted science. Let us therefore brace ourselves to our duties, and so bear ourselves that, if the British Empire and its Commonwealth last for a thousand years, men will still say, 'This was their finest hour.'

Winston Churchill, later Sir Winston Churchill, was renowned for his command of the English language, his turn of phrase, and his ability to galvanise Parliament and the British citizens into action. He was an inspiration for the personnel to produce the best engine they could for the war effort and Rolls-Royce had, as it turned out, a very important part to play in winning the war. Ernest Hives, later 1st Baron Hives, initially Head of the Experimental Department, became General Works Manager in 1936 just as the Merlin had had its first flight in the Hawker Hurricane.

Sensing the future, he split off the aero engine division of Rolls-Royce and prepared the company for mass-production of the Merlin engine. This was before war had been declared. It was a decision of immeasurable strategic importance, resulting in Rolls-Royce producing, with sub contracts, approximately 150,000 Merlins for the war. Hives' foresight ensured that, if at all possible, there would be sufficient units available to meet the immediate demand. Air Chief Marshal Wilfrid Freeman was a, if not the, most important leader in the rearmament of the RAF in the pre-war years. In 1936, he held the position of RAF Air Member for Research and Development, in which capacity he made the decisions on which aircraft to rearm with and, later, in 1938, he was made responsible for production. He ordered the Merlin engine Hawker Hurricane, Supermarine Spitfire, de Havilland Mosquito and Avro Lancaster among other aircraft types. One stroke of brilliance was to re-engine the North American P-51 Mustang with the Merlin engine. Freeman said of Hives: 'He cares for nothing except the defeat of Germany and he does all his work to that end.' Post-war, Hives would become Rolls-Royce's chairman of the board.

A quote from the *Daily Telegraph* of 6 November 2015, gives an idea of the man's presence. It was from Derby's Deputy Mayor, Councillor Mark Tittley, himself a former Rolls-Royce employee: 'When entering the front door of the Marble Hall to receive a warm greeting from the smartly dressed commissionaire, although the bronze busts of Charles Rolls and Henry Royce were quite rightly present on either side of the hall, it was, for me, always one other personality that dominated the scene; that of Lord Hives.'

Ernest Hives was with Rolls-Royce for most of his working life. Starting with the company as driver/(car) tester well before the First World War, he was promoted to General Works Manager in the mid-1930s (then a very powerful position in the company) and, subsequently, he became Chairman and Managing Director post-war, when he was also raised to the peerage. In many ways, Lord Hives, or to use his old Rolls-Royce reference 'Hs', remains one of this country's unsung heroes and, from what one has heard about him or read, he probably preferred it that way.

It was the strategic imagination and leadership of Tommy Sopwith at Hawkers, R. J. Mitchell at Supermarine and Hs at Rolls-Royce, that saved the country from defeat in the Battle of Britain (along with the great skill of our pilots and the workforces who built the products they flew). These three men had the insight and drive to ensure that, in spite of the apparent lack of interest of British politicians in the mid- to late-1930s, who were wrongly following a policy of appeasement with Hitler's Germany, the development and building of the Hurricane, Spitfire and the Merlin engine were pursued anyway. Thank goodness, they did, not just for this country but also for the whole of the democratic world.

It was from his office located by the boardroom at the Main Works and, of course, in the boardroom itself, that Hs directed the development and production of the Merlin and subsequent Griffon engines, both before and during the war; not only at Derby but also as they came 'on-line' during the war in the 'shadow factories' at Crewe and Glasgow. Lord Hives demonstrated what great leadership is really about, not just in terms of making great strategic decisions at the right time, as with the decision to build the Merlin pre-war or to take the company into gas turbine design and production in the commercial aerospace sector post-war, but also in terms of how he interacted with the workforce. As an example of the latter, I have been told how Hs would appear out of the blue on a factory floor or in a technical office area working on a difficult problem.

A common phrase heard is that the Hurricane and Spitfire, Britain's most famous Second World War fighter duo, were designed 'around the Merlin' engine. What would this engine do for these and other Merlin-powered aircraft? In one word – everything. The Merlin engine would propel these fighting machines faster and higher than aircraft had gone before and, with their armament and the aerodynamics of their wings and fuselage, allow the skill of the pilot to turn these aircraft into formidable lethal weapons. Of course, their fighting superiority would not last forever as they would be superseded, in spite of the continuing development of the Merlin, it too had a physical limit to how much power it could generate. The Rolls-Royce Griffon and the Rolls-Royce Avon would take over, with the Avon heralding the jet age. The Merlin had fulfilled its requirements.

The Merlin was a Rolls-Royce private venture engine that had come to the attention of the government and would subsequently end up coming to the attention of Sydney Camm. Who was Camm and why was he so important to the Merlin engine? He did not work for Rolls-Royce and yet he was instrumental in the start of the Merlin's road to fame by it subsequently powering some very important and famous aircraft – the first production fighter being Camm's own design.

Camm was born at the end of the nineteenth century in Windsor, Berkshire. Technical genes ran in the family and his brother Frederick, a technical author, created *Practical Wireless* magazine. Camm left school at fifteen and began a career as an apprentice carpenter. Simultaneously he developed an interest in aeronautics and began building and selling model aircraft to people who could afford them, namely the students at Eton College. Five years later, in 1912, he was one of the founding members of the Windsor Model Aeroplane Club.

It was a natural progression for him to enter the world of aircraft manufacture with the Martinsyde Aircraft Company, which was located at the Brooklands racing circuit in Weybridge, Surrey. The Brooklands circuit was the first-purpose built, banked racing circuit in the world. It was opened in June 1907 and hosted a procession of cars, one driven by Rolls himself. Two years later, it became one of the first British airfields. Alliott Verdon-Roe (AVRO) had carried out towed flight trials there the previous year.

It soon became a major aircraft manufacturing and flying training site associated with names such as Bristol, Vickers, Sopwith, Bleriot and the company that Camm had joined, Martinsyde. The race track was closed during the war and aviation took over the entire facility. Camm was quickly promoted to the drawing office and spent the war years honing his skills. Martinsyde went bankrupt in 1921 and Camm joined the Handasyde Aircraft Company, which created the Handasyde monoplane. His career continued to advance, and he changed employment to the Hawker Aircraft Company in Kingston upon Thames, where he was now a senior draughtsman. His first design was the Hawker Cygnet, an ultralight biplane which first flew in 1924. The success of this aircraft led to him being appointed chief designer in 1925.

As the transition from biplane to monoplane continued, in 1925 he developed, with a Hawker associate, a type of tubular metal construction using cheaper and simpler joints than the previous welded structures. This led to a string of designs including the Tomtit, Hornbill, Nimrod, Hart and Fury. The 1925 Rolls-Royce Condor IV-engined Hornbill never went into production, the 1928 Tomtit was a radial engine training aircraft, the

1928 Rolls-Royce Kestrel-engined Hart was a bomber aircraft, the 1931 Rolls-Royce Kestrel-engined Fury was a fighter and the 1933 Rolls-Royce Kestrel-engined Nimrod was a carrier-based fighter. All these aircraft were biplanes and were the predecessors of the monoplane Hurricane. In the 1930s his designs were prevalent in the RAF until other designers emerged with successful aircraft.

This brilliant aeronautical engineer, like so many others of his calibre, was not the easiest person to work with. Recollections of some of his workers, such as Sir Robert Lickley who later became chief engineer at Fairey Aircraft, were that he was a demanding taskmaster and woe betide anyone who was not willing to go above and beyond the normal amount of effort to achieve his aims. The impending threat of a European conflagration spurred Camm on to take his previous experience from biplane design and apply it to the emerging monoplane design and construction. The result, among others, was the first Merlin-powered fighter, the pre-war Hawker Hurricane.

Camm did well with the Hawker Aircraft Company; but what was its origins and ambitions when Camm joined it in the 1923? The company was named after an Australian, Harry George Hawker.

Hawker was born in 1889 in Moorabbin, Victoria, Australia. By all accounts Hawker was no academic in his early school years but he certainly had mechanical aptitude and interest. In 1900, at eleven years of age, he was helping to build engines in a Melbourne garage. It is unbelievable in the twenty-first century that a child would be so employed back then. Hawker then moved on to the Tarrant Motor Company, where he qualified as a mechanic. Inquisitive by nature, he attended the first public demonstration of powered flight at Diggers Rest near Melbourne. The next year found Hawker, at the age of twenty-two, in England following his dream to enter the new exciting world of aviation.

Work was not as plentiful as he expected and without a local reference, his future employment prospects looked dim. However, he managed to get a position with Commer, a commercial vehicle company, which led to further positions with two motor vehicle companies, the Mercedes Company and the Austro-Daimler Company. It is ironic that they were both German companies, which would bear the brunt of Hawker aircraft wrath in the war that was to follow.

Having spent some time visiting at Brooklands, which was the centre for both aircraft and car development in England, he was fortunate to get a position with the growing young concern called the Sopwith Aviation

Company. It had a small workforce, but Hawker was one of them, he had gained entry to the aviation industry.

Thomas Sopwith was just a year older than Hawker and had developed an interest in motorcycles and ice hockey, winning a gold medal at the European Championship. In 1906, he had his first balloon ascent with Charles Rolls; it certainly was a small world of qualified aviators in those years. Sopwith learned to fly in 1910 and in November was awarded Royal Aero Club Aviation Certificate No. 31, flying a Howard Wright 1910 biplane. In December, he won a financial prize for the longest flight to the Continent in a British aeroplane and with the winnings he set up the Sopwith School of Flying at Brooklands in 1912.

Hawker had scarcely started work with Sopwith as a mechanic when he started flying lessons as a pupil with his employer. He qualified as a pilot, certificate No. 297, in September and very quickly became a flying instructor to recoup the price of his flying lessons. Hawker was a natural pilot and wanted to continue his aviation adventure by becoming involved in flying competitions. His attempt at the longevity competition succeeded with a flight of more than eight hours to win the British Empire Michelin Cup and the prize money. Hawker's next competition target was the altitude record held by another famous name in aviation, Geoffrey de Havilland. He succeeded in breaking that record and several others also.

By December 1912 Sopwith Aviation Company had its first military order and had moved premises to Kingston-on-Thames. Hawker was the company's chief engineer and test pilot during the war years. In 1920, the Sopwith Company fell on hard times and, together with Thomas Sopwith and others, Hawker bought the assets of the company and formed H. G. Hawker Engineering. Unfortunately, a year later Hawker was killed in a Nieuport Goshawk at Hendon Airport, London. Camm joined the firm, by now named Hawker Aircraft Company, two years after the founder's death and the Hawker named lived on with the establishment of Hawker Aircraft Limited in 1933, the same year as the Merlin was on the drawing board and a prototype was being built.

The Air Ministry was the department of the UK Government responsible for the Royal Air Force that was established towards the end of the First World War in early 1918. The Air Ministry survived until 1964. Politically it was the responsibility of the Secretary of State for Air. It started its life with the creation of the Air Committee within weeks of the Royal Flying Corps being formed in 1912. The Air Committee lacked any executive power, as it was an intermediary between the Admiralty and the War Office in aviation matters and could only make recommendations. After the start of the First World War in 1914, when it was really needed as the Army and Navy's aviation programmes continued to separate, it never met again.

By 1916 it was obvious that more co-ordination was needed between the Royal Flying Corps and the Royal Naval Air Service to be two effective fighting forces. A Joint War Air Committee was formed early in 1916 to bring about the change. In fact, it was just as ineffective as the previous committee and the chairman resigned. In July, another attempt was made by forming the Air Board. Once again, the needs and aspirations of the Navy and Army could not be reconciled. In 1917, under the new leadership of Lord Cowdray, some progress was made but, to cut a long story short, it took the intervention of the Prime Minister Lloyd George and one of the members of the British Imperial War Cabinet, General Jan Smuts, to recommend that a new ministry be formed, with Lord Rothermere as Air Minister and President of the Air Council. It met in January 1918. A year later it moved to Adastral House, Kingsway, in London.

The various Secretaries of State had their own visions of the development of aviation. The difference of emphasis went from airships to overseas flying; Imperial Airways was established, linking the Empire. The Royal Air Force College Cranwell and the University Air Squadrons were established at this time. The Supermarine aircraft entering the Schneider Trophy races were flown by RAF pilots and the teams had limited support from the Air Ministry.

It was with the Air Ministry that the various aircraft manufacturers and their designers had to work in the inter-war years. Specifications were issued by the Air Ministry and the onus was on the companies to design an aircraft that would be deemed suitable to meet, or exceed, the requirements. These design proposals would be assessed and a prototype might be ordered for further evaluation. Sometimes, very rarely, the companies would go ahead with a private venture design and build the aircraft and hope it impressed the Air Ministry enough to issue a specification based on the aircraft. Strangely enough, the Air Ministry would assign its name, if it was indeed ordered. The ordering process was called ITP, Intention to Proceed. The maximum price was fixed but the actual price was up for negotiation as government pressure was applied.

The specification issued by the Air Ministry originated with a description of what the role of the aircraft would be, the Operational Requirement (OR). The OR.32 description would be further refined in the specification, such as a single-engine fighter with four guns, Specification F.6/37. This specification system was in place from 1920 until 1949. It encompassed the time of the Air Ministry, the Ministry of Aircraft Production (MAP) and the Ministry of Supply.

Perhaps this is a good time to digress slightly to discuss the Ministry of Aircraft Production. Churchill, when he became Prime Minister, wanted to remove the MAP from the Air Ministry, but who to head it? He remembered the Minister of Munitions from the First World War, Lord Beaverbrook, and appointed him for his 'vital and vibrant energy'. Beaverbrook made an immediate important decision to incorporate the Civilian Repair Organisation and RAF repair under his Ministry. Perhaps indicative of Beaverbrook's approach to the MAP was his reaction to the exodus of workers from the factories when the air raid siren sounded. He wanted them to remain working until the enemy aircraft arrived instead of rushing home. In Reading he managed to wangle some cement and had shelters built at the factory. In his office he had a slogan, 'Organisation is the enemy of Improvisation'. Beaverbrook worked his subordinates very hard, going over and over every way of increasing production. He led by example and often toured factories. The results of the week's production were examined every Saturday. No effort was to be spared, nor any idea regarded as too ludicrous to improve production. One of Beaverbrook's actions was to release German Prisoners of War who had experience with extruding aluminium and place them in aircraft factories. Some sabotage? Perhaps, but in the main, production went up.

Unorthodox to say the least, he often tendered his resignation to get the Prime Minister to intervene and get the necessary supplies or action that he needed. The Beaverbrook's Pots and Pans Appeal stated: 'We will turn your pots and pans into Spitfires and Hurricanes.' This netted tons of aluminium but the real effect was that he was very popular among the housewives and established a loyalty base. This was followed by the Spitfire Fund. Production was his life at that time. He made some errors of judgement: delaying a vital reconnaissance aircraft was one. However, Air Chief Marshal Dowding said of Beaverbrook: 'We had the organisation, we had the men, and we had the spirit which could bring us victory in the air, but we had not the supply of machines necessary to withstand the drain of continuous battle. Lord Beaverbrook gave us those machines, and I do not believe that I exaggerate when I say that no other man in England could have done so.' In Britain's hour of need he got the job done.

The Air Ministry specification was always in a standard format, although not necessarily issued in sequence. The first letter designated the aircraft type, F for Fighter and B for Bomber are examples, and the number grouping would indicate the issued specification sequence, followed by the year it was issued. A few other examples would be the letter N for Naval, T for Training, C for Cargo and E for Experimental. Two or more manufacturers could submit designs, or prototypes, for the same specification and this would be referred to as Avro B.8/31 and Handley Page B.8/31, for example.

A very important specification was F.7/30 to replace the Gloster Gauntlet fighter. The year 1930 marked the very early stages of the biplane transition to the monoplane. The Air Ministry specified a fighter capable of 250mph (400kmh) with four machine guns, low landing speed and short landing run, high initial climb rate, high manoeuvrability and good pilot visibility. Hawker submitted the biplane PV3 with a Rolls-Royce Goshawk steam-cooled evaporative engine but was not successful. Gloster was awarded the contract and produced the Gloster Gladiator, interestingly a private venture, which was the last RAF biplane fighter ordered as it was already made obsolete by the new monoplane designs being submitted. Two of the rejected monoplane designs submitted for F.7/30 were the Supermarine Type 224 and the Bristol Type 133. It would be a crucial delay of four years until 1934 before another fighter OR, OR.15 F.5/34, was issued. This resulted in four companies – Bristol, Gloster, Martin Baker and Vickers – building monoplane aircraft but none went into production.

Camm had designed the Hawker Fury I & II in the early 1930s and by using the same fabric-covered tubular construction for overall strength, slightly tapering the leading and trailing edges of the wings, and replacing the Rolls-Royce Goshawk with the emerging heavier Rolls-Royce Merlin engine, Camm was able to move the centre radiator aft and install a retractable undercarriage. Speeds now could be anticipated in the 300mph (480kmh) range and higher. Camm was assisted by Roland (Roy) Chaplin who, in addition to assisting in the design work, on occasion it is reported helped to smooth the ruffled feathers of the support staff, namely the secretaries. Camm drove his staff as hard as he drove himself.

It was not the leaders or the rank and file of the aviation industry who dragged and pushed the industry into the monoplane age, it was the middle-rank RAF officers and civil servants who saw the future of the monoplane and encouraged Camm, and Mitchell of Spitfire fame, to pursue their ideas. This encouragement resulted in major changes in thinking, such as relocating the guns to the wings and increasing their number. The designers had discussions with the Directorate of Technical

Design in 1933 but it was not until late 1934 that finally, OR.16 F.36/34, which specified a 'High Speed Monoplane Single Seat Fighter', resulted in the Hawker aircraft that would go into production. A mock-up was presented to the Air Ministry in January 1935.

Sir George Edwards, former Chairman of the British Aircraft Corporation, reflected on the effect of the Schneider Trophy and said: 'There can be no doubt that the boldness of these designs and the passion for engineering detail which they displayed made a profound impact on aeronautical design and set the scene for the successful generation of British fighters which were so decisive in saving Britain from defeat in later years. If the industry had been limited during these inter-war years to design studies alone and not been able to translate ideas into hardware by actually building aeroplanes, it is certain that such successful fighters as the Spitfire and Hurricane would not have emerged.'

The prototype, K5083, was manufactured at the Hawker factory in Kingston-on-Thames and was moved in pieces to Brooklands aerodrome in October for assembly and subsequently test flown by George Bulman on 6 November 1935. Bulman had flown Sopwith Camels in the Royal Flying Corps during the First World War and had been awarded the Military Cross for his role in the Battle of Courtrai. He was a test pilot in the newly formed RAF when he transferred to Hawker in 1925. Including the Hurricane, he made the first flights of nine Hawker aircraft. Bulman is reputed to have leapt out of the cockpit after the Hurricane's first flight and said to Camm: 'It's a piece of cake. I could even teach you to fly her in half an hour, Sydney.' On 9 February 1945 Bulman's only child, his son Raymond, was killed flying a 605 Squadron Merlin-powered de Havilland Mosquito. The Bulman family's contribution to the war effort included the ultimate sacrifice, like so many others.

Merlin cooling had been a problem and necessitated a large radiator, not conducive to clean aerodynamic lines. Camm solved this by having a smooth air flow to an unobtrusive duct underneath the fuselage, which directed the air to the radiator. The initial wings were fabric-covered and were filled with ballast to account for the missing armament. Two problems were noted after the test flight: the canopy creaked and flexed, and the engine temperature increased during taxiing and lowering the flaps, which disturbed the airflow.

Various sources called the Hawker aircraft the 'interceptor monoplane' or 'experimental fast aeroplane'. The Merlin had test-flown only six months earlier for the first time and it was still having teething problems as it gained test flight hours. The Merlin 'C' engine had been installed in the first Hawker prototype and in fact it had not yet achieved full flight certification. The 'C' produced 1,029bhp and weighed 1,180lbs (535kg). In a previous attempt, it had also failed certification and was in effect 'a work in progress'. The 'work' even included a partial engine failure, luckily with no damage to the prototype. In December, the Merlin C was issued a provisional airworthiness certificate, not a great start for a legend. However, the plan had already dictated that the Merlin C would not be used for production but it would be succeeded by the newer and improved Merlin F, later known as the Merlin I.

By March 1936 the prototype was sent for initial service evaluation at the Aircraft & Armament Experimental Establishment (A&AEE) at Martlesham Heath. Coincidentally, on 5 March, the Supermarine Spitfire went airborne for the first time at Eastleigh airport, Hampshire. The airframe gained a reputation for reliability and A&AEE reported ease of handling and good control at all speeds. The wide stance undercarriage gave good ground handling and rough field operational capability, but the thick wing section had its advantages and disadvantages. The advantages were that the wing was robust and had the room for fuel, armament and undercarriage, while the disadvantage was that the Hurricane would always be aerodynamically limited, although the increasing power and reliability of the Merlin would lessen that disadvantage.

The Merlin was still in its infancy and was being forced to grow up fast with the threat of war. Leaks, distortions and cracking dogged the initial engines and would require fixing. There was a further complication in that the engines might not be ready for the first aircraft coming off the production line. As previously mentioned, one of the fixes was to create the engine in three major components, the crankcase and two cylinder blocks. It is simple to write here but you can imagine the panic and activity at Rolls-Royce until the solution was found?

Hives had to make the executive decision to restart the new engine production. Thanks to the inevitable delays of Hurricane production, according to Major Bulman at the Air Ministry: 'Rolls were able to regain their substantial lead in Merlin deliveries to meet the aircraft output. But it was a harassing few months, peace mercifully still prevailing.' Other teething problems that had to be dealt with were supercharger bearing failure, broken valve springs and boost control failures. It was not an easy infancy.

However, despite these problems, the performance figures were a triumph: 315mph (507kmh) in level flight and a climb to 15,000 feet (4572 metres) in less than 6 minutes. The Merlin and Hurricane were coming of age and by June 1936 an initial order for 600 aircraft was on the Hawker books. Twenty of this first order were diverted to the Royal Canadian Air Force. The build-up of monoplane fighters to protect Britain had begun. Would the four-year delay in the transition to monoplane fighters affect the outcome of the future war or would human tenacity, skill, dedication to the war effort and guts make up the difference? The decision makers in 1936 now fully realised the implications of the evolving political situation and hoped, above all hope, that it would. On 27 June 1936, the prototype was designated the Hawker Hurricane, a name that would go down in Britain's aviation history. Each Hurricane would rely on a Rolls-Royce Merlin engine.

The Hurricane proved amenable to mass production and improvements occurred regularly. The original two-bladed wooden propeller was replaced with a de Havilland metal, three-bladed, two-pitch propeller that resulted in a better rate of climb and better fuel economy. However, the first aircraft to receive the production Merlin 1, 172 of them, was the Fairey Battle bomber, which flew in March 1936. The switch to the Merlin II for fighter production delayed the programme, both because of the engine change and the resulting airframe modifications, until 12 October 1937, when the first production Hurricane I flew, a short and frantic two years before the outbreak of the Second World War. The delay was calculated at six months, the bad news, but the good news was that the final product was well worth the wait. Would that delay play a major role in the final result of the war? Finally, Britain had a basic monoplane fighter aircraft that would initially equip the RAF squadrons and had the potential for improvement both in its engine, the Merlin, and its airframe.

By 1938 the British government responded to the increased demand by sanctioning licensed limited production of the Hurricane in foreign countries, Canada being one of them. The Canadian Hurricanes were powered by the US Packard-built Merlins. There was a need was to ensure supply of Hurricanes and that meant keeping the factories well away from the reach of the enemy bombers. The Canada Car and Foundry (CanCar) at Fort William, now Thunder Bay, Ontario, certainly met that criteria. Although endorsed by the Air Ministry, it was almost an entirely commercially sponsored enterprise. Initially, microfilm drawings and a few examples of the Hurricane were sent to Canada to provide patterns for production facilities. Later, twenty-two examples were sent over as initial development aircraft to launch production training schedules. This was so successful that 160 Canadian-built Hurricane Is participated in the

Battle of Britain. The chief engineer was Elsie MacGill and she became famous and was known as the 'Queen of the Hurricanes'.

MacGill's contribution to the Allied victory is another one of the many 'people' stories that fade into history but should never be forgotten. Her story is a human triumph linked to and with the Hurricane. Even contracting polio before her graduation in electrical engineering did not slow her down, she learned to walk supported by two (probably well-engineered) metal canes. She was the first woman in Canada to graduate as an electrical engineer. After further studies at the University of Michigan, MacGill was awarded a Master's degree in Aeronautical Engineering in 1929, certainly the first woman in North America and possibly the world to hold such a distinction. She completed her doctoral studies at the Michigan Institute of Technology by writing magazine articles about aircraft and flying.

In 1934, she started work with Fairchild Aircraft in Longueuil, Quebec, as an assistant Aeronautical Engineer, eventually becoming the Chief Aeronautical Engineer with CanCar. MacGill designed, built and tested a biplane training aircraft called the Maple Leaf Trainer II. The Hawker Hurricane contract was awarded to CanCar and MacGill was responsible for the workforce, which grew to 4,500 workers, half of them women. She streamlined factory production and designed modifications, de-icing controls and skis replacing the wheels on the landing gear to allow the Hurricane to operate in the harsh Canadian winter climate. At her passing in 1980, Shirley Allen, a Canadian member of the Ninety-Nines organisation of women aviators, said of her: 'She had a brilliant mind and was recognised as an outstanding Canadian woman. Neither gender nor disability prevented her from using her talents to serve her community and country.'

Other countries that were involved in the export of the Hurricane during the threat of invasion were South Africa, Rumania, Poland and Turkey. Yugoslavia and Belgium were also awarded production licences for the Hurricane but the war intervened before more than a few could be built. Hence the decision to have the aircraft built in Canada, far from the action. Approximately ten per cent of all Hurricanes were built in Canada.

The Hurricane was a less complicated aircraft to manufacture than the Spitfire. It required thirty per cent fewer man hours to produce, and was cheaper to manufacture because of its relatively simple construction and the use of established production methods. I have always thought of the Hurricane as being the family car and the Spitfire as the sports car. By December 1938 the first four Hurricane Is entered service with the RAF No. 111 Squadron stationed at RAF Northolt, now part of north-west Greater London. The Merlin, in late 1938, was still having teething problems as recounted by Squadron Leader D. H. Clarke, in Robert Jackson's *Hawker Hurricane*: '… Far worse than this was the Merlin engine, which used to cut out for no reason at all – generally at the worst possible moment.'

This was quickly followed by sixteen aircraft to No. 111 in February 1939 and aircraft to RAF No. 3 Squadron at Biggin Hill. Hawker production reached a total of nearly 500 aircraft by the outbreak of hostilities in September 1939. The Merlin and Hurricane were designed and built for a reason, to shoot down enemy aircraft. On 21 October 1939, No. 46 RAF Squadron, flying out of the North Coates satellite field, intercepted and shot down four Heinkel HE 115B floatplanes.

The original Hurricane I had been upgraded to a later model Hurricane I, which featured various improvements. Rolls-Royce had developed the Merlin III engine and fitted it with a British Rotol, developed from the Hele-Shaw system of the 1920s, constant speed propeller; ejector exhaust stacks provided extra thrust; fabric-covered wings were replaced by stressed metal-covered wings, and the 'ring and bead' gun sight was replaced by a standard 'reflector gun sight'. All of these increased the performance of the Hurricane I. The communication package was improved with an Identification Friend or Foe system and radios to communicate to the ground intercept controllers. More important to the pilots was the retrofit of the fuel tank in front of the pilot with fire-resistant material to help prevent burn injuries. Further flight testing had revealed a slow spin recovery, which was cured by the addition of a small ventral fin to improve the air flow around the bottom of the rudder.

Bulman was assisted in later years in flight testing Hurricanes, both in development and checking production aircraft prior to squadron delivery. Bill Humble was a production line test pilot for eight years before becoming Sales Manager. He recounted that 'after eight years my nerves were pretty well shot and it (the new office position) came as a great relief.' In 1931 Philip Lucas joined Hawker as a test pilot and fifteen years later, he was the General Manager. Lucas had demonstrated and delivered the Hawker Hart and Hawker Fury throughout the world in the early 1930s. He was awarded the George Medal for his performance during a Typhoon flight test. It read: 'Test Pilot Lucas displayed great courage and presence of mind during a test flight and, by his skill and coolness, saved an aircraft from destruction.'

As with the introduction of any new aircraft there were many modifications done, which resulted in many aircraft versions with slight differences. This was further amplified by the advent of the so-called 'phoney war' as Britain scrambled to build up its fighter strength. A major headache for the ground crew was an aeroplane that looked like a standard Hurricane until the cowling came off for closer inspection! Adding to the confusion was that some of the Hurricane Is were being upgraded to II status. The Battles of France and Britain will be dealt with in the next chapter but in May 1940 Hurricane Is were dispatched to Norway.

The Hurricane and Spitfire are synonymous with the Battle of Britain, but this was a defensive campaign. The war would not be won by defensive measures alone, just contained, a stalemate at best. The Hurricane would later go on the day fighter offensive for a short time before assuming its next important roles, the night fighter interceptor, the fighter bomber and the intruder role. These were the offensive times for the Hurricane and with its Merlin engine it was taking the fight to the enemy and eventual victory. Rolls-Royce was again rising to the challenge and produced a more powerful Merlin XX engine for those new roles.

The war would become global and the Hurricane with its Merlin engine would find itself in the far corners of the British Empire and other countries. The Hurricanes were dispatched down the Mediterranean to fight in Greece and Malta and by a circuitous route to North Africa and the Middle East. They could be found in the Far East in India, Burma (now Myanmar), Ceylon (now Sri Lanka) and Singapore. The Hurricanes would carry the fight from the southern hemisphere to the Arctic Circle in Russia. The Royce-designed Merlin piston engine would cope with temperatures ranging from +40°C to -40°C, quite an accomplishment, quite an engine. It would also go to sea to protect the North Atlantic convoys, Britain's vital supply line.

Notwithstanding the number of Hurricanes built, the types of Merlin engines installed in them were few in retrospect. The Hurricane Mk I had the Merlin Mk II and III engines. The Hurricane Mk I was further improved by the replacement of the wooden propeller by a constant speed metal propeller. The Hurricane Mk IIA, IIB and IIC had the benefit of the Merlin XX engine with its two-speed supercharger. The engine was slightly longer, which made the aircraft slightly more stable. The Hurricane Mk IV had the Merlin 24 or 27 engines. The Hurricane Mk V never went into production, but it had the Merlin 32 installed and had a four-bladed propeller. The Hurricane Mk X series were all Canadian-built and had the American-built Packard Merlin 28 and 29 installed – more about Packard in Chapter 10.

In *The Hurricane Story* by Paul Gallico, Camm is quoted as saying: 'When the design of the Hurricane had gone beyond the point of no return I suddenly had a foreboding that it would be no good. I was always a worrier. Sopwith, Sigrist, Bulman and I had all decided that strength should be an important factor in this ship, but I always had one regret. I wished the wing had been thinner ... Yet, if we had not gone ahead we should have had nothing when we went to war with Germany.'

There are plenty of stories about the Hurricane but very little mention specifically of the installed Merlin. That would seem to the author to imply that it performed as requested, although the fighter pilot would always want more power, but power coupled with manoeuvrability. The consensus seems to be that the Hurricane was a stable gun platform of concentrated firepower with better visibility for attacking bomber aircraft.

Group Captain Peter Townsend, RAF 43 Squadron, RAF Acklington, said 'the Hurricane was more manoeuvrable at its own speed (than the Spitfire) and undoubtedly the better gun platform.' It was acknowledged that the Supermarine Spitfire was more agile and had better performance, although, because of its narrow track landing gear, it was trickier to land in a crosswind or rough terrain.

Importantly, the Hurricane could turn tighter than the Spitfire and its opponent, the Messerschmitt ME109, and that took power, Merlin power. This turning ability was enhanced, according to Maurice Allward in his book *Hurricane Special*, 'by the pilots using the engine boost override continually during interceptions, a practice which the Merlin engine withstood remarkable well.'

Michael Bywater, in an article in the 8 July 2010 edition of *The Independent* newspaper, garnered the following quotations about the Hurricane: 'It took a lot of muscle to haul her round the sky. It became a good friend from the start, and I loved it more and more. It had the wonderful ability to absorb battle damage. The plain one, the dumpy one (as compared to the Spitfire).' This was reflected in the success of the Spitfire Fund, whereas the Hurricane Fund went nowhere.

Wing Commander Jack Rose said: 'I became more and more impressed by the robust qualities of the Hurricane.' Perhaps the greatest tribute to the Merlin as told to Bywater was: 'I never had to worry about the Merlin engine ... you're flying very low, a couple of trees high off the ground, and you do want it to keep going.'

But what of the other half of the Battle of Britain victors, the Supermarine Spitfire? Whose brain created the concept for this sleek fighter, which would follow the Hurricane on the road to fame? What ideas and aircraft had gone before to sow the seed of this successful fighter? Both aircraft were emerging in the forefront of monoplane development. What was the Spitfire's pedigree?

Two years after Camm was born, another brilliant aeronautical engineer was born near Stoke-on-Trent, Staffordshire, and his name was Reginald Mitchell. Like Camm, as a child he also had an interest in model aeroplanes, designing, making and flying them. After leaving grammar school in Stoke-on-Trent, he started an apprenticeship at sixteen with Kerr Stuart & Co. of Fenton, a locomotive engineering works. At the conclusion of his apprenticeship, he was promoted to the drawing office and furthered his study of and interest in engineering by taking engineering drawing, mechanics and mathematics at night school. In 1917, after six years of study, he joined Supermarine Aviation Works at Southampton, Hampshire. Within two years he was named Chief Designer at age twenty-four, a very quick advancement by Supermarine, which recognised his talents. In his relatively short career he designed twenty-four aircraft, a prolific output. Based in Southampton, Supermarine was primarily a seaplane manufacturer. Mitchell's designs included the Supermarine Sea Eagle, Sea King, Walrus, and Stranraer. The author had the pleasure of flying with some Queen Charlotte Airlines Stranraer pilots early in his career and enjoyed listening to their stories of flying the Stranraer along the coast of British Columbia, Canada.

Mitchell did not restrict himself to seaplanes and also designed light aircraft, fighters and bombers. Notwithstanding these aircraft of varying success, he is principally remembered initially for his racing seaplanes taking part in the Schneider Trophy competition. Mitchell was determined not only to win the race for Britain but to win it three times in succession to secure the Trophy for the nation in perpetuity. As previously described in Chapter 2, the Supermarine S.5, S.6 and S.6B, designed by Mitchell, by now the Technical Director, won the trophy for Britain in 1927, 1929 and 1931. The last two victories were achieved using the Rolls-Royce R engine. The S.6B was the ultimate aerodynamic design of the racing seaplane and this experience contributed greatly to the future design of the Supermarine Spitfire. He was awarded a CBE for his work in connection with the Schneider Trophy.

Mitchell's next design of importance, the Type 224, was a failure, but lessons learned from it produced the Type 300, a private venture by the Supermarine Company. Just prior to the Type 224 flying in early 1934, Mitchell was diagnosed with cancer. He continued to work despite his illness and designed the Type 317, a four-engine heavy bomber. In addition to his work on the drawing board, he received his pilot's licence in July 1934. By 1937 his illness had returned, and he was unable to work any longer. He did observe the prototype Type 300, subsequently called the Spitfire, being tested but never saw it enter full production; he passed away in June 1937. What a legacy he had left Britain in her time of need.

Mitchell did well with the Supermarine Company but what was its origins and the company's future when Mitchell joined it in 1917? The company was formed by quite a 'character'. Noel Pemberton-Billing was born in 1881 in Hampstead, a suburb of North London. He left home at thirteen and went to South Africa to fight in the Boer War. He returned to England in 1903, the same year that the Wright Brothers managed the first sustained powered flight of a heavier-than-air aircraft. Pemberton-Billing became a lawyer, but his interest was in the new aviation industry, he could see its potential.

Several attempts at establishing aerodromes and flying fields failed. Ever the adventurer, he supposedly bet Frederick Handley Page that he could earn his pilot's licence within twenty-four hours of sitting in an aircraft. He won the bet and was issued pilot's licence No. 683. In 1913, he opened an aircraft business, Pemberton-Billing Limited, and registered the telegraphic address as Supermarine, Southampton, at Oakbank Wharf, Woolston. The company struggled financially in the flying boat business and, after Billing departed for the Royal Naval Air Service in 1914, he sold the business two years later to his Works Manager, Hubert Scott-Paine, who renamed it Supermarine Aviation Works Limited.

The company continued with a stream of floatplanes, flying boats, amphibians, Trophy Race seaplanes for both military and civilian applications. The Type 224 was the forerunner of the Spitfire. The low-wing monoplane unfortunately suffered from problems with the evaporative cooled Rolls-Royce Goshawk engine. The company was taken over by Vickers-Armstrong in 1928 and one of its conditions was that Mitchell stay on for five years. After Mitchell's death in 1937, the further development of the Spitfire was under the leadership of Joseph Smith, who became the Chief Designer. He had been Chief Draughtsman at Vickers-Armstrong since 1926 and was heavily involved with the Spitfire design working under Mitchell.

Mitchell, like Camm, responded to the government specification F.7/30 but he submitted a monoplane, Type 224, design unlike the Hawker

PV3 biplane design. So already Mitchell was pushing the boundaries of design and engineering and was not afraid to submit these proposals for important government contracts, thus pushing Supermarine to the forefront of the British aircraft manufacturing scene. Frank Barnwell and Archibald Russell of the Bristol Aeroplane Company also submitted an unsuccessful monoplane design, the Type 133, which flew on 8 June 1934. The Type 133 had gone one better than the Supermarine submission, it had retractable undercarriage. However, it was all to no avail, because as mentioned previously the government awarded the contract to an already obsolete biplane, the Gloster Gladiator.

The Type 224, which first flew on 19 February 1934, was an inverted gull-wing monoplane with a fixed spatted undercarriage using the Goshawk II evaporative cooling system engine. The condensers for the cooling system were in the leading edge of the wing with the collector tanks in the undercarriage legs. This necessitated the armament being placed either side of the cockpit and in the undercarriage fairings to meet the specification requirement of four machine guns. The gull-wing configuration was used to shorten the drag-producing non-retractable undercarriage legs. A lack of directional stability caused Mitchell to enlarge the fin. The aircraft had disappointing performance and he was looking at designing major changes when the specification, in effect, itself became obsolete. The armament, now eight guns, and performance requirements had to be changed, so a new specification was drawn up instead of modifying the old one.

Mitchell had learned a lot from the Type 224 and immediately designed several optional designs drawing more on his Schneider Trophy experience. The obsolete drag-producing fixed undercarriage was removed and replaced with an outward-retractable undercarriage, and split flaps were added to a reduced wing span. These changes were not enough of an improvement, so the team went back to the drawing board. The cockpit was enclosed and with the new Rolls-Royce PV12 engine now available, a new smaller and thinner wing could be used, as the Goshawk condensers were no longer needed. The Canadian aeronautical engineer, Beverley Shenstone, was responsible for this high performance elliptical – more correctly ellipsoid – wing, which would give the Type 300, later called the Spitfire, its distinctive shape and made it easily identifiable from the ground. The elliptical wing gave low wing loading and the long chord at the join to the fuselage gave it strength.

Shenstone graduated as Canada's first-ever Master of Aeronautics from the University of Toronto. Ten years before the war, he was working at the Junkers factory in Dessau and soaring at the famous Wasserkuppe gliding facility in Germany. The Wasserkuppe, a large plateau, is the highest elevation at 3,120 feet or 950 metres in the Rhön Mountains. Every German aeronautical engineer of note spent time there in the 1920s and 1930s putting their flight theories to the test. In 1930, Shenstone found himself at the leading edge of aerodynamic theory and practice. He worked in Germany with such luminaries as Alexander Lippisch, of delta wing fame, and Ludwig Prandtl, a calculus and boundary layer specialist. Through his associations in Europe he was exposed to the work of the Russian pioneer Nikolai Zhukovsky and his research on elliptical wings. Shenstone, by now Mitchell's Chief Aerodynamicist, said that Mitchell jokingly said to him: 'I don't give a bugger whether it's elliptical or not, as long as it covers the guns.'

When the new specification F.5/34 had appeared, Mitchell was ready for it, this time there would be fewer design errors and more performance. Ten months after the Type 224's first flight, the Type 300's construction began. He had worked closely with Rolls-Royce before with the 'R' engine for the S.6 and S.6B. The PV12 had not flown yet and one of its problems was overheating, which was solved by ethyl-glycol coolant. Finally, at the end of 1935, the engine and airframe came together. Unlike the Hurricane – which used many jigs, tools and procedures from previous aircraft construction – the Spitfire was an entirely new approach to construction methods, such as heavy gauge metal wing cover forward of the main spar, in effect a solid box. It was a major challenge for Supermarine and the Mitchell team with war looming on the horizon. It should be mentioned that Dowding of the Air Council made the far-reaching decision in the early 1930s that wood was not to be used as the structural base of combat aircraft.

In the spring of 1936 the prototype was ready for its first flight. There is some doubt as to the exact date, 5 or 6 March 1936. The following is from Jeffrey Quill's book *Spitfire: A Test Pilot's Story* as he comments on the date of the first flight; 'On 6 March I flew the Falcon from Brooklands to Martlesham to take Mutt Summers (Vickers Chief Test Pilot) from there to Eastleigh. Popular folklore has it that the first flight of the Spitfire took place on the 5th of March 1936, but I flew Mutt to Eastleigh for the particular purpose of making that first flight on 6 March.' After the flight, Quill returned Summers to Brooklands. Coincidentally, the other prototype, Hurricane K5083, was sitting in a nearby hangar at Brooklands. Both aircraft at this time were mere shadows of the fighters they would become. On 7 March Hitler's forces occupied the demilitarised zone in the Rhineland.

On 26 March Jeffrey Quill was the second person to fly the Spitfire: 'I primed the Merlin engine carefully and it started first time.' Quill had served in the RAF and had done such a good job commanding the RAF Meteorological Flight that he was awarded the Air Force Cross. In 1935, he had left the RAF to work with Summers as his assistant at Vickers flight testing. Quill subsequently flew every Spitfire mark produced. In August 1940, he rejoined the RAF for a short time to experience combat in the Spitfire and promptly shot down a Messerschmitt Bf 109. In 1943, he once again left Vickers to fly in the Fleet Air Arm to experience carrier operations and have input into the improvement of the Seafire.

After Mitchell's death in 1937, the further development of the Spitfire was under the leadership of Joseph Smith, who became the Chief Designer. He had been Chief Draughtsman at Vickers-Armstrong since 1926 and had been heavily involved with the Spitfire design, working under Mitchell.

Production began in March 1937 with Supermarine initially constructing the fuselage and with the other components widely subcontracted. For example, some wings were built by Pobjoy Airmotors and Aircraft Ltd, and Singer Motors Ltd built the engine mountings. There were early manufacturing problems as it was a complicated all-metal aircraft, and these delayed delivery until August 1938. Meanwhile, on 22 March 1937, the prototype K5054 was undergoing high 'G' turns and looping trials at Martlesham Heath to test new elevator gearing. The oil pressure dropped to zero and the engine began to run rough, so Summers cut the engine and skilfully belly-landed the aircraft in a field. This enabled the damaged engine to be examined and subsequent Merlins were modified to prevent oil starvation during high 'G' manoeuvres. By this time the Merlin on the prototype had new exhaust manifolds, and the aircaft had a castoring tailwheel, reduced rudder size and a raised panel on top of the wing surface to indicate if the flaps were down.

In 1938 aircraft K9834 was being prepared for an attempt on the world air speed record. The powerplant was a specially strengthened Merlin III using 100 octane fuel. The engine had a four-bladed fixed pitch propeller, which gave 2,160hp. Before this attempt at the record, the bar had been raised twice by a Heinkel and Messerschmitt to a speed of 50mph (81kmh) above the calculated speed for the Spitfire. The attempt was abandoned.

I do not think at this time any further comment on the Merlin is required except this quote from Quill in his book *Spitfire: A Test Pilot's Story:* 'At the end of it all I felt a very friendly disposition towards the new Merlin engine… I learned to be meticulous in the matter of correct engine handling at all times, and although I never hesitated to run a Merlin to the absolute limit of its capabilities, I was careful never to exceed those limits except when unavoidable… The Merlin really was the pilot's friend.' Quill tested the first production Spitfire on 15 May 1938. It was apparent then that the Woolston factory alone would not be able to keep up with the contracted delivery schedule, especially as Vickers was initially loath to release the blueprints to outside contractors. This did cause a delay in deliveries. It would take until 1940 for the Castle Bromwich shadow factory to get up to speed under the guidance of Vickers.

The first production Spitfire to be delivered to a squadron was K9789 to No. 19 Squadron RAF, at RAF Duxford, on 4 August 1938. Squadron Leader, later Air Commodore, Henry Cozens took it flying on 11 August, a scant nine months before hostilities started. This was the first Spitfire delivered and it would take time to build up the numerical squadron strength and then to train the pilots in their new aircraft, a monoplane, and in their new role. Twelve days later, the second aircraft was delivered to the squadron. This trickle would obviously become a flow but for now at least it had begun. It would take nearly to the year's end for the squadron to get its full complement of sixteen aircraft. A total of 20,351 Spitfires were eventually built, plus 2,408 Seafires.

K9789 was flown at maximum effort to accumulate flying hours to test for weaknesses, and some corrective action was taken to fix discovered defects. The flying hours were also used to formulate a rudimentary maintenance schedule. Cozens recalled in Alfred Price's *The Spitfire*: 'But that was not to say she was perfect. For one thing, the engines of these first Spitfires were difficult to start: the low-geared electric starters rotated the propeller blades so slowly that when a cylinder fired there was usually insufficient push to flick the engine round to fire the next; there would be a 'puff' noise, then the propeller would resume turning on the starter. Also, the early Merlin engines leaked oil terribly; it would run down from the engine, down the fuselage and finally got blown away somewhere near the tail wheel.' The installation of a faster starter quickly cured the starting problem, but the leaking engine took longer. The propeller was changed and experimented with to find the perfect operational balance with the Merlin. The original two-bladed wooden propeller was good at the top end of the performance envelope but not at the bottom end, take-off being an example. A three-blade fine pitch propeller was tried, followed by a two-pitch three-blade propeller, fine for take-off but coarse for operations.

The technical solution was the constant speed propeller in 1940, which automatically gave the correct pitch settings for all airspeeds.

H. Taylor recounted humorously in a 1946 edition of *Flight*, regarding the jamming of the undercarriage selector: '... it was necessary to take the weight off the retracted legs (undercarriage) if any further landing progress was to be made. Since, of course, the only way of doing this was to invert the aircraft, Spitfires might occasionally be seen on their backs during the circuit.' This would definitely qualify as a defect to be corrected!

The Spitfire Mk I had the Merlin II, 1,310bhp at 9,000ft (2,743m), or Merlin III, 1,440bhp at 5,500ft (1,676m), engine installed. Initially the Spitfire had eight .303 in Browning machine guns in the wings with 300 rounds of ammunition. It had 85 gallons of fuel and a top speed of 364mph (586kmh). It had retractable landing lamps underneath the wings for night operations. In June 1939 four Browning guns were removed and replaced with two Hispano cannons to make the Spitfire MkIB. This was not a success due to cannon stoppages and the conversion ended with just thirty aircraft. By September 1939, nine RAF squadrons were up to Spitfire strength, Nos. 19, 41, 54, 65, 66, 72, 74, 602 and 611.

Alex Henshaw was the Vickers test pilot at Castle Bromwich, Birmingham. He says of the Merlin engine in his book *Sigh for a Merlin*: 'Inevitably the pressures of wartime demanded shortcuts in production, as with the piston failures. These in no way detract from the superb design and fantastic performance of these (Merlin and Griffon) great engines.' The Castle Bromwich wartime statistics show 127 forced landings for 33,918 test flights! Henshaw said in Quill's *Spitfire A Test Pilot's Story* that '... when I look back on the intensity of the flying in those years, and the tremendous bashing handed out to various Merlins in the course of rigorous performance testing, I marvel at the relatively few times the engine let me down and look back with gratitude.'

The Merlin and Spitfire were designed and developed for a reason, the reason turned up on 16 October 1939 in the Firth of Forth. Nos. 602 and 603 RAF Squadrons, RAF Renfrew and RAF Turnhouse respectively, engaged a force of German bombers attacking shipping. Two Junkers 88s were shot down. The defence of Britain had begun. It would be a year until the Battle of Britain came to an end; the island had already been successfully defended by the Merlin engine.

Pilot Officer David Crook of 609 Squadron at RAF Drem, Scotland, in *Spitfire – Living Legend* by John Dibbs and Tony Holmes recounts that on his first flight in a Spitfire '...and then (I) opened up to full throttle. The effect took my breath away. The engine opened up with a great smooth roar, the Spitfire leapt forward like a bullet and tore madly across the aerodrome, and before I had realised quite what had happened I was in the air.'

At the start of hostilities in September 1939, the RAF was equipped with the initial series of Hurricane and Spitfire aircraft. Production was ramping up and the improvements in aircraft and engine design had already started. The Merlin was going to war, first to defend the homeland and next, when that was secure, to attack the enemy.

The Merlin would be called upon to perform many different roles. The short aggressive aerobatics of fighter combat, the seven-hour high-altitude bomber raids in a multi-engine aircraft, and everything in between. The next five-and-a-half years would test the entire country, including the aviation industry and all its suppliers, and none more so than Rolls-Royce and its Merlin engine.

In all the published articles and books about the Hurricane and Spitfire, there is very little mention of the engine. Speeds, manoeuvrability, ease of landing on rough surfaces, stable gun platform, turning radius, etcetera, are all mentioned. It seems to be limited to what version of the Merlin was installed in what Mark. It would seem that it was expected that Rolls-Royce would continue to improve the Merlin and that it could be relied upon to perform to the usual high standards. The manufacturers could rely on their powerplant and concentrate on improving the performance and firepower of the aircraft. Rolls-Royce had fulfilled their wartime commitment and it would not be 'for the lack of an engine that the war was lost'.

The 'phoney war' was about to come to an end in early 1940. How would these new monoplane fighters match up against the might of the Axis fighters and bombers? Would they have the necessary manoeuvrability and firepower to overcome the enemy with, initially, relatively inexperienced young pilots? Britain would soon find out, initially with the Battle of France followed closely by the ultimate test of man and machine, the Battle of Britain.

5

The Battle of France &
The Battle of Britain
1939–40

The end of the First World War in November 1918 of course came as a great relief to the global populace. There never had been such devastation and loss of life witnessed in wartime, it was 'the war to end all wars'. This 'Great War' was followed closely by a worldwide flu epidemic, which, in fact, killed more people worldwide than the war did; the world was reeling. This period was followed by a worldwide recession in the late 1920s. Germany had in addition the emasculating Treaty of Versailles to contend with. It was a peace treaty signed between the Allies and Germany to prevent a war from happening again.

Coming to power in 1933, Hitler capitalised on the binding restrictions of the treaty and the poor economic conditions of the time. Hitler felt that Germany was hemmed in and needed 'living space'. He immediately tested the willingness of the Allies to monitor and enforce the treaty. Quietly rebuilding the air force, the Luftwaffe, and navy, the Kriegsmarine, under several guises, the scene was set for expansionism. By 1936 he had remilitarised the Rhineland and this was followed by the annexation of Austria and questionably, to prevent even greater expansion, Germany had been handed part of Czechoslovakia by the French and British at the Munich conference in September 1938. Appeasement did not work, Hitler took the rest of Czechoslovakia in March 1939. The invasion of Poland on 1 September 1939 was too much for the Allies and finally on 3 September 1939 they declared war on Germany. The unthinkable had happened in twenty years; another, as it turned out, world war had begun.

BATTLE OF FRANCE
Prior to the Second World War, England and France had made an agreement that in event of war, the RAF light bomber force

would relocate to bases in France to attack Germany. The Advanced Air Striking Force (AASF) was formed in August 1939. The AASF initially consisted of eight Fairey Battle squadrons, two Bristol Blenheim squadrons and two Hawker Hurricane squadrons, Nos. 1 and 73, which soon would be supplemented by No. 501 Auxiliary Air Force. No. 1 Squadron previously flew Hawker Furies at RAF Tangmere and No. 73 Squadron flew Gloster Gladiators at RAF Digby. The squadrons had all formed during the build-up in 1938 and had just enough time to become familiar with the aircraft and develop tactics.

The Fairey Battles, with Merlin engines, and Bristol Blenheims of the AASF initially were not allowed bombing sorties from French soil for fear of reprisals, but eventually they hampered the enemy advance with bombing. The Hurricanes accompanied and protected these light bombers. On 2 September the light bombers deployed to airfields in the Rheims area, the Merlin engine was now in Europe prepared for war. On 8 September, sixteen No.1 Squadron Hurricanes, fabric-covered MkIs with wooden propellers, deployed to Le Havre moving via Cherbourg to Vassincourt on 15 September. No.73 Squadron deployed to Caen on 10 September.

The Hurricane was ideally suited to the undrained, soft ground, temporary airfields. The wide stance and sturdy undercarriage coped with the challenging operations from these inadequately prepared landing strips. The squadrons performed routine patrols in the border area and occasionally saw Luftwaffe reconnaissance aircraft in the vicinity of the front. The enemy would take cover by diving into cloud. Initially the squadrons moved quite a bit and the accommodations varied from wooden huts to lavish digs. One thing remained the same, the muddy fields. It was indeed, for now, a phoney war.

Identification was very important at this time as the ground troops were not familiar with the Hurricane. Each aircraft had red, white and blue stripes on the tail. Squadron markings on the fuselage consisted of the squadron code, VY for No. 85 Squadron, the roundel and the actual aircraft, for example F for Freddie. There was also a unit marking – a white hexagon for No. 85 Squadron.

First blood would go to No. 1 Squadron, flying out of Vassincourt, as recounted in Francis K. Mason's book *The Hawker Hurricane*. An excerpt from the official diary states: 'Vassincourt. 30th October 1939. Local patrol maintained throughout the day. At 1430 hrs three enemy aircraft were seen at a high altitude over the aerodrome. The aerodrome defence section immediately took off in pursuit and one of the raiders was overtaken at 18,000 feet, 10 miles west of Toul, and shot down by Plt. Off. P. Mould (in Hurricane *L1842*). The other two escaped in cloud. The enemy shot down was a Dornier Do 17 and appeared to have been taken by surprise as no evasive tactics were employed and no fire was encountered by Plt. Off. Mould.' Shortly after, on 2 November, Flying Officer E. J. Cain of No. 73 Squadron, in a de Havilland variable pitch Hurricane, shot down a Dornier Do 17 at a record height of 27,000 feet; the AASF's war had begun. Early in 1940, No. 1 Squadron received Merlin III Hurricanes with the Rotol constant-speed propellers.

The French aero engine industry was unable to meet the home demand and was looking around for other manufacturers to fill the void. The Merlin was establishing a reputation and Fordair, a subsidiary of Ford of America, was requested by the French Air Ministry to negotiate an agreement to manufacture the Merlin in France. Monsieur Dollfuss, the managing director, indicated to Rolls-Royce that they were not really interested in building complete engines at this time and that perhaps the request by the Air Ministry was more a political gesture than real. However, Rolls-Royce had to treat the request seriously, as all-out war was becoming more likely and perhaps it could turn out to be a good financial proposition.

The British Air Ministry kept a close account of manufacturing rights overseas and established certain conditions. The first was that the rights be restricted to the Merlin III only. There would be a lump sum paid by the French Government, plus a royalty per engine. Rolls-Royce would provide a technical information exchange between Rolls-Royce and Fordair, to keep them fully aware of possible developments. Finally, the Air Ministry would receive a proportion of the royalty paid to Rolls-Royce. Engineers arrived at Derby from the parent company in the USA and special machine tools were ordered from Ford, Dearborn, Michigan. To facilitate the start-up, complete sets of Merlin parts would be sent to France in September. However, already struggling as a viable manufacturing agreement, it came to an ignominious end in August. Henry Ford had no faith in France and Britain withstanding the might of Germany, so he made the decision to make his company neutral. Engineers were withdrawn, parts orders cancelled and eventually the whole deal collapsed with the French factory unable to go it alone, and perhaps not being given full support by the French Government. Although not manufactured in France, Merlin engines were installed in two French aeroplanes, the Amiot bomber and the Dewoitine fighter. Perhaps the failure was a blessing in disguise, in that Rolls-Royce did not have to divert energy and resources to a French factory, which, as later events would reveal, would have fallen into enemy hands.

The British Expeditionary Force (BEF) had been established in 1938 as war became inevitable. Immediately after the declaration of war in 1939, the BEF deployed to the north-east of France, bordering Belgium. Hurricane squadrons, Nos. 85 and 87, provided fighter cover for the BEF. A period of time elapsed as each side contemplated military moves, as in a deadly chess game, and built up its resources. The build-up, in fact, was more along the lines of a deficiency make-up. The inter-war years had initially lulled Britain militarily, probably as a result of mental and financial exhaustion, and politically as it coped with the Depression. In spite of this, by the middle of October, the BEF consisted of 152,000 soldiers and 25,000 vehicles and by March 1940, the complement of men had doubled in size.

The BEF dug in on the border alongside the French First Army. This was still the 'phoney war', as not much activity was apparent, with the BEF or on the home front in Britain. The war became 'real' in April 1940 when Germany invaded Denmark and Norway, followed by the invasion of France, Belgium, Holland and Luxembourg on 10 May 1940. Some Allied success occurred with the Dyle Plan in Belgium, but it was soon discovered that the main attack had occurred farther south through the Ardennes and a retreat ensued. By 10 May there were seven Hurricane squadrons in the BEF Air Component, Nos. 3, 79, 85, 87, 504, 607 and 615. Peter Jacobs in his book *Hawker Hurricane* states that on 10 May, 'Large sections of the French, Dutch and Belgian air forces had already been destroyed. More than 200 Hurricane sorties had been flown; forty-two enemy aircraft were confirmed as destroyed, for the loss of seven Hurricanes.'

Particularly rewarding days for 501 Squadron were 11 and 12 May; the enemy aircraft destroyed on those days were six and twelve respectively. The Hurricane's role was changing from bomber escort to troop support as the situation worsened. On 13 May, thirty-two Hurricanes were sent to France to bolster the effort. That is when Air Chief Marshal Dowding made the decision to stop sending any more resources overseas; he anticipated, rightly, that they would be needed at home. On 12 May the first two Victoria Crosses for bravery were posthumously awarded to Flying Officer Donald Garland and observer Sgt. Thomas Gray in a Merlin-powered Fairey Battle for their attack on the Veldwezelt Bridge in Belgium.

The Air Force priority was now the Second Army at the Sedan bridges on the Meuse River and many escorted bomber sorties were flown there on 14 May. With German fighters outnumbering the Allied fighters 3:1, the bomber loss rate was extremely high at approximately 60 per cent. It is generally accepted by historians that the loss of the Sedan bridges sealed the fate of Belgium and France. By 17 May, although destroying sixty enemy aircraft, attrition had reduced the air component to the equivalent of an amalgamated three operational squadrons in the north. The German advance to the English Channel at Abbeville on 21 May outflanked the Maginot Line and would trap and encircle half a million Allied troops.

On 21 May it was over; the remaining pilots and aircraft were ordered to return to England. Every possible Hurricane was flown home; as long as it flew, it went. The recovered aircraft, as well as home-based Spitfires not previously used in France, were ordered to protect the retreating BEF troops. By 26 May small craft were assembling at Dunkirk on the French coast to evacuate the BEF and Allied armies. In Operation Dynamo, that day 3,000 sorties were flown with 200 enemy aircraft destroyed at the terrible cost of approximately 100 pilots.

On 27 May No. 501 Squadron had one of the greatest air battles fought by Hurricanes. The squadron was operating out of a forward strip at Boos, near Rouen. The eventual tally was eleven Heinkel IIIKs destroyed, three Heinkel IIIs and one Messerschmitt Bf 110 possibly destroyed, and all squadron pilots returned safely, quite a feat.

On 28 May two new RAF Biggin Hill squadrons, Nos. 213 and 242, started operations over the beachheads in France. They encountered Heinkels, Junkers and Messerschmitts that were attacking the troops. The evacuation lasted nine days and the squadrons destroyed twenty-six aircraft for the loss of nine Hurricanes; it seemed that the Hurricanes and the Merlin were working well. All squadrons were going airborne five or six times a day to cover the evacuation, a herculean effort.

Paul Gallico in his book *The Hurricane Story* wrote: 'Squadron-Leader Harry Broadhurst fought the Germans with his Hurricanes with good effect, but at such high altitudes that the grousing ground troops did not even know they were being given air cover, leading to Churchill's remarks about the victory inside the deliverance at Dunkirk gained by the Air Force. Over the brown beaches of the French coast, six ME 109s jumped him and he ended up "jinking like a mad waterbug" only a few feet above the sea with his Hurricane depreciating rapidly. His aeroplane started to take on the appearance of a sieve with holes shot in his controls and his canopy shattered.

With a bullet through the Merlin engine its heart kept beating in spite of losing oil and he managed to crash-land at RAF Northolt and walk away to fight another day. The aircraft was written off but Broadhurst had the aircraft restored, that is another story, and flew it when he was a Wing-Commander as his private aircraft. Reunited with his old Merlin friend.'

On 3 June, the remnants of the AASF withdrew to six strips south-west of Paris. The German army overran the Dunkirk docks on 4 June. The Hurricanes were defending the troops at Nantes, Brest and St Nazaire and by 18 June most of the Hurricanes were on home soil or had been abandoned and destroyed on French soil. The RAF had forty aces – pilots with five or more confirmed kills – during the Battle of France.

The Merlin had contributed to the effort by getting the aircraft to, and sustaining the aircraft in, the aerial combat over the evacuation beaches to allow the pilot to fight his air battle. The evacuation continued farther down the French coast until 25 June and later still, August, from the Mediterranean ports. In less than a month France had fallen, with Allied troops leaving the country and it would be four years until they returned, and the Merlin would once again be heard on the ground in France.

In the Battle of France the BEF Air Component lost seventy-five out of 261 Hurricanes and 120 were unserviceable or lacked fuel. Only sixty-six were repatriated. The AASF lost sixty-six Hurricanes. In all, 477 fighters were lost, but the real story was the loss of pilots. The fighters were being replaced at 400 per month but in mid-1940 the loss of experienced pilots was only being replaced by a trickle of newly trained, inexperienced pilots. This was the weak link, the greatest factor that would test Britain in its greatest air challenge yet, in the summer of 1940. Could the Commonwealth's youth not only rise to this challenge but excel enough to defeat the impending air armada approaching Britain's shores?

BATTLE OF BRITAIN

On 10 May 1940, the same day as the start of the Battle of France, Winston Churchill was appointed Prime Minister of Great Britain. On 18 June 1940, he made his famous speech to the House of Commons, already mentioned in Chapter 4, in which he said the Battle of Britain was about to begin and the nation needed to brace itself to do its duty so that 'if the British Empire and its Commonwealth last for a thousand years, men will still say, "This was their finest hour."'

Within three weeks of the 'Fall of France' the Nazi war machine, confident and visualising Europe secure, made plans to invade Great Britain and remove the thorn in its side. That thorn would prove to be part of a pretty sturdy briar bush and eventually its blossoms would spread the scent of freedom throughout Europe. Most (though not all) Germans and Luftwaffe personnel had no idea what they had started in July 1940 and by the end of October 1940 the resolve of the British people and her Allies changed the direction of the war. Britain now had the opportunity to further develop her offensive campaign. The Allies called this time in 1940 the Battle of Britain and eventually it was won. Easy to state but Britain's survival was a close call. A tactical mistake by Hitler and the German High Command, perhaps made because of arrogance and an instant angry reaction to the retaliatory bombing of Berlin, gave Britain unexpected relief and a chance to rebuild.

The Merlin would be called on to perform as never before in its history. Scrambled day and night at a moment's notice, it would take off at maximum power from the airfields of Britain and claw its way into the sky looking for ever more altitude. Within minutes it would be engaging the enemy and, in the race to achieve supremacy, it could be at emergency boost power. Twisting and turning, with positive and negative 'G' loading, at warm or cold altitudes and varying speeds, from the stall to maximum diving speed, the Merlin would be expected to perform in all these conditions to keep its pilot alive. and in addition to these mechanical challenges it would be expected to do this, in some cases, up to five times a day. Fuel, oil and glycol coolant were its lifeblood and its doctors were the ground crew who ensured that it was in the best of mechanical health. Roll-Royce had done its work and now it was up to the pilot to make the most use of the engine and the aeroplane he was flying to give him supremacy in the air.

However, it would take even more than the lone fighter to stop the enemy. It would take a system, facilitated by an invention, to create a defence network to make the best use of the Merlin and its aeroplanes. That invention was radar and the system was set up by Air Chief Marshal Hugh Dowding, Commander of Fighter Command. Together these two examples of human ingenuity enabled small and large groups of fighters to provide the best defence against the approaching bombers and escorting fighters. Through foresight, its limited capability in the late 1930s would grow to a formidable force during the war. The 'Dowding' system would enable the fighters to win the Battle of Britain during the summer and autumn of 1940 and turn the tide of the war.

A Scotsman, James Clerk Maxwell, published a paper on the dynamic theory of the electromagnetic field in 1864. Maxwell wrote, 'The agreement of the results seems to show that light and magnetism are affections of the same substance, and that light is an electromagnetic disturbance propagated through the field according to electromagnetic laws.'

Heinrich Rudolf Hertz was a German physicist who, 20 years later, built laboratory experiments to explain Maxwell's electromagnetic theory of light. Hertz measured Maxwell's waves and proved that the velocity of radio waves was equal to the velocity of light. Hertz was the first to satisfactorily demonstrate the existence of electromagnetic waves by building an apparatus to produce and detect Very High Frequency and Ultra High Frequency radio waves. He also discovered and demonstrated the reflective properties of radio energy. It would be another 20 years, in 1904, before Germany's Christian Hülsmeyer demonstrated that by using radio waves he could detect the presence of metallic objects. He used the example of a ship in dense fog. Radar was born, but nobody was interested, such is the life of an inventor.

Finally, in 1935, another Scotsman, Robert Watson-Watt, arguably 'The Father of Radar', demonstrated to the British Air Ministry the ability to detect an aircraft by radio methods. On 17 June 1935, the Daventry Experiment proved that it was possible to detect signals from the local BBC radio transmitter reflecting off a circling RAF bomber 13km away. By the end of 1935, Watson-Watt had a patent for radar, initially called Radio Detection Finding by the British, and he increased the detection range to more than 100km and installed five radars, Chain Home (CH) stations, to protect the approaches to London in 1937.

The stations consisted of massive metal transmitting towers and wooden (to avoid interference) receiving towers. The CH system was capable of providing range and height information of approaching aircraft, which was passed on by phone line to a central command center. Not easily destroyed, the superstructure being a difficult target for the dive bombers, but fairly easily rebuilt, these stations were the backbone of the radar defence of Britain during the beginning of the Second World War. The initial systems were at first manually operated, then were improved with the installation of automatic features as the system constantly improved. Height finding, and identification friend or foe, were also added during development. Eventually, the basic CH systems gave way to a plan position indicator system in 1942.

Ken Peacock, a Canadian radar mechanic posted during the war to the Broad Bay CH radar site near Stornoway, Western Isles of Scotland, recalled: 'The west coast stations were different. We had two 240 feet (73 metre) towers with the MB2 transmitter and RF7 receiver. On my first night there, I had to climb the tower. The power amplifiers could put out 650,000 watts. We operated on the 12-metre wavelength. It was fixed direction and we covered the North Minch area. There was also an identification friend or foe transmitter for aircraft, if there was no blip (friendly identification) on the radar return we would phone Stornoway, our filter station.'

By the outbreak of war there were twenty-one operational coastal CH stations, mainly on the east and south coasts. The secret to the success of the system was the dissemination of the data received from the CH stations. It was the first integrated controlled interception network in the world and was operational in September 1939. This would be a network that required intensive labour, using the technology available at the time. The stations consisted of two sites, the transmitting site and the receiving site. Each site would have an operator and the receiving site would have an assistant operator and six assistants operating the plotters, calculators and communication lines, and thousands of miles of dedicated private lines. Multiple crews were needed to give twenty-four hour coverage, plus there would be additional service and support personnel. The Women's Auxiliary Air Force played an important part in supplying trained personnel for these positions.

The CH system was a very basic system for raid assessment, often relying on the skill of the operator as to the size and height of the incoming raid. It used a fixed omnidirectional antenna and once the target had passed, it was no longer of any use for that target. It was not uncommon for CH stations to pick up enemy formations forming over France. The raid information from the coastal CH stations was relayed by phone to the 'filter room' at the Fighter Command central processing location at RAF Bentley Priory. These CH reports would be combined with Royal Observer Corps (ROC) visual reports and the raid would be assigned a track identified by number. This track information would be phoned to Group Headquarters to assign fighter resources. The Group would assign squadrons and in turn pass the information on to the affected sector, which was in direct contact with the fighters to scramble, or, if airborne, intercept, the enemy aircraft. It took an enormous number of staff, working twenty-four hours a day, to make the system work.

The CH system was not the first defensive system for Britain. The London Air Defence Area had been put in place during the First World War. A plotting board was established at Horse Guards in London to receive visual reports in a central area and then send the information outwards to the relevant response units, anti-aircraft artillery, searchlights and fighter aircraft.

The CH station antennae were difficult to hit, they were just a framework, and could be repaired. Sometimes a double bluff would be played by not immediately repairing the damage to give the impression that it was not very important. They, in fact, were one of the reasons for the Allied fighter pilots' timely encounters with and eventual defeat of the Luftwaffe in the skies over Britain. CH was just one cog in the machine that made up the RAF Fighter Command, which was active as a single entity for seven years between 1936 and 1943. The Command was divided into Groups, with 11 Group taking the brunt of the attack in 1940 in London and south-east England, 10 Group was responsible for south-west England, 12 Group for Midlands and East Anglia and 13 Group for the north of England and Scotland.

Fighter Command Headquarters and Group HQs were built in bombproof bunkers, away from airfields. Sector Control Centres were above-ground buildings, protected by an earth bank or blast wall. These centres all had communication phone lines buried very deeply to help prevent bomb damage. Of course, all these centres would have to have twenty-four hour coverage from the support and supply personnel. It took an army of people to get the Merlin and its aircraft with its energy and firepower to the scene of the combat as quickly and as efficiently as possible. It is still debated today whether, if it had not been for the Dowding system timely in place in 1939, the Battle of Britain would have been lost. Winston Churchill said: 'All the ascendancy of the Hurricanes and Spitfires would have been fruitless but for this (Dowding-created) system which had been devised and built before the war. It had been shaped and refined in constant action, and all was now fused together into a most elaborate instrument of war, the like of which existed nowhere in the world.'

It required the guidance of Hugh Dowding, Air Officer Commander-in-Chief Fighter Command since 1936, to bring the system to operational status. The CH and ROC flow of information was overwhelming at times and Dowding realised he also needed a way of distinguishing the enemy from the defence fighters, hence the identification friend or foe transponder installed in the defending fighters. Dowding built the filtering system that ensured that only the information required by the pilot was forwarded to them. The plots, small wooden blocks with tags, indicated whether they were the enemy or friendly. The blocks were all colour-coded to the sector clock colour, so that the position of aircraft could be seen easily as an old or recent plot. The squadron assigned to the intercept was also displayed on the track block.

The sectors had a commander, who was a pilot who had retired or was on medical leave, thus adding authority to the pilot-to-pilot communication. The idea was to prevent the fighter pilots hunting for targets and perhaps wasting resources, instead obeying the instructions of the sector controller who was aware of where the enemy aircraft were and, most importantly, to avoid friendly-fire incidents where the friendly aircraft were. The Sector Commander was also in contact with the barrage balloon and anti-aircraft batteries.

A 'tote board' kept track of the squadron status and displayed it as Released: not available, Available: airborne in 20 minutes, Readiness: airborne in 5 minutes, Standby: pilots in cockpit, airborne in 2 minutes, Airborne and moving into position, Enemy sighted, Ordered to land, Landed and refuelling/rearming. The WAAF kept the tote and weather boards up-to-date. This tote board concept, now computerised, is still used today by the North American Aerospace Defence System and other defence systems throughout the world. It is a simple but effective way to provide the commander with all the necessary information to use the resources at his disposal. The system proved that the working combination of military attributes far exceeded the sum of these attributes working alone. It took the individual available forces and by combining them multiplied the resulting total force.

High-altitude bomber formations travelling at faster speeds were too much for the old system. The result was standing patrols and fighter sweeps producing no enemy sightings, far less engagement. Interception, not success, rates were 30 per cent. It would take massive numbers of fighters to patrol an area and this was not possible. This concept gave rise to the grim philosophy that 'the bomber will always get through'.

The CH stations and the Dowding system changed all that. Timely and accurate information on the raids was reaching the fighter squadrons in enough time to find and attack the enemy. Interception rates rose to 75 per cent and continued to rise as the operators became more experienced. A defence wall grew around Britain but it did have a few cracks in it. Sometimes the flow of information to Fighter Command HQ overwhelmed the operation and some information flow was stopped, disregarded or got lost. The fighters' low-powered radios were inadequate for the system; they had only one voice channel, set to a specific squadron frequency before takeoff, which was susceptible to interference. Dowding was removed from his position in Fighter Command in November 1940 due to the 'Big Wing' controversy, still debated by historians, but his system worked when it was really needed in the 'summer of the fighter'.

'The Few', as the pilots were referred to by Churchill, could rely on the Merlin and the rest was up to them. 'The Few' were mainly British pilots but also included pilots from the Commonwealth and elsewhere. Australia, Canada, New Zealand, South Africa, what was then Rhodesia, Belgium, Poland, Czechoslovakia and France were among the countries represented. They also included pilots from neutral countries who felt that they had to contribute to what they saw as a just cause, such as Ireland and the USA. Two of the Group Commanders were from overseas, Group 11's Air Vice-Marshal Keith Park was from New Zealand and Group 10's Air Vice-Marshal Sir Quintin Brand was from South Africa.

The RAF was reeling after the loss of 477 fighters during the Battle of France and now in the summer of 1940 had to replace those fighters and the lost pilots, who had been killed, were missing, were Prisoners of War, or injured, with newly trained airmen. Fifty-two fighter squadrons were calculated as being necessary for the defence of Britain, on 7 June there were only thirty-six squadrons that were combat-ready. The standard Hawker Hurricane I at this time, in early summer 1940, was the Merlin III engine driving a Rotol or de Havilland variable-pitch three-bladed propeller; the standard Supermarine Spitfire Mk I had the Merlin III engine driving a three-bladed constant speed propeller. The Hurricane, by dint of its construction, proved to be easier to repair after battle damage. A total of 527 Hurricanes and 321 Spitfires, approximately 600 operationally serviceable, would face three times as many Luftwaffe bombers and fighters. As previously mentioned, 11 Group in the south-east of England would bear the brunt of the attacks.

It is beyond the scope and not the intent of this book to give an account of the Battle of Britain, but all the leading aces were in Merlin-powered Hurricanes or Spitfires, it was indeed the engine that won this battle. Only eight per cent of all Allied pilots qualified as an ace, having shot down five or more aircraft. These pilots far exceeded the minimum requirement during the battle of 1940.

Kills: enemy aircraft confirmed destroyed   **KIA: Later killed in action**

P/O Eric Lock	41 Squadron RAF Hornchurch	26 kills KIA
Sqn/Ldr Archie McKellar	605 Squadron RAF Drem, Croydon	19 kills KIA
Sgt James Lacey	501 Squadron RAF Middle Wallop, Gravesend	18 kills
Sgt Josef Frantisek	303 Squadron RAF Northolt	17 kills KIA
F/O Brian Carbury	603 Squadron RAF Turnhouse, Hornchurch	15 ⅓ kills
F/O Witold Urbanowicz	145/303 Squadron RAF Northolt	15 kills
P/O Colin Gray	54 Squadron RAF Hornchurch	14 ½ kills
P/o Bob Doe	234/238 Squadron RAF St Eval, Middle Wallop	14 kills
F/Lt Paterson Hughes	234 Squadron RAF St Eval, Middle Wallop	14 kills KIA
Sqn/Ldr Michael Crossley	32 Squadron RAF Biggin Hill, Acklington	14 kills

The Battle of Britain went through phases as it developed. The initial phase was the German attack post the Battle of France in July on coastal targets and shipping operating in the English Channel. Then on 13 August, *Adlertag* ('Eagle Day'), the focus became the airfields and communication centres. On 31 August Fighter Command suffered its worst losses of the campaign so far, as it defended the RAF fighter resources. This phase continued for a little over three weeks and then on 7 September, believing the RAF was on its last legs and incensed at the retaliatory attack on 25 August on Berlin, the attack by the Luftwaffe switched to the city of London, the 'Blitz'. This was devastating for the citizens of London but it gave the defences, the RAF bases, communication centres and radar installations, time to recover. On 15 September Fighter Command repelled the largest number of aircraft ever in a Luftwaffe attack. The enemy aircraft suffered unsustainable losses. The writing, to German disbelief, was on the wall and, although fighting continued, within a few days Hitler cancelled

'Operation Sealion', the planned invasion of Britain. Upon reflection, it is interesting that the Germans consistently underestimated the size and power of the RAF and that the British consistently overestimated the size and power of the Luftwaffe.

Not all Germans – and certainly not all those who were doing the fighting and the dying – were shocked. *Staffelkapitän* Gunther Scholz of 7/JG 54 recalls in Patrick Eriksson's book *Alarmstart*: 'Göring claimed (in summer 1940) that factories had been destroyed, that the balance of fighters remaining to the RAF ... was minimal and thus not much more could be left in their arsenal. We had to laugh silently to ourselves about this nonsense and (*Generalleutnant* Adolf) Galland dared to state our experiences clearly and distinctly. He fell into Göring's disfavour for a while after this.'

In researching this book, it was impossible to come up with a definitive number for the losses and claims of the Luftwaffe and the RAF, there were too many sources with different statistics. In the confusion of an air battle, especially at night, confirming the number of enemy aircraft hit, damaged or lost was a challenge at best. At worst, further postwar calculations revealed inaccuracies from all combatants, RAF and Luftwaffe, that were further enhanced by the propaganda machine. There was no way that the British public would be exposed to negative statistics during their epic struggle. Suffice to say that a lot of young men, 'The Few', in Merlin-powered fighters, gave their lives and, supported by the Dowding system, had won the Battle of Britain but not the war. The tide had turned and now it was time for the Merlin to take the bombs to the enemy.

PART 3

THE MERLIN TAKES THE WAR TO THE ENEMY 1940–45

The Bomber Offensive from Britain

6

Fighter-Bomber:
De Havilland Mosquito
1940

What goes faster than a Merlin? Two Merlins attached to a very light aeroplane. The de Havilland Aircraft Company had been working on a light, twin-engine, mostly wooden, aeroplane to compete in the MacRobertson Air Race, a long-distance multi-stage journey from the United Kingdom to Australia, which was being planned for October 1934 to celebrate the 100th anniversary of the founding of the Australian State of Victoria.

Geoffrey de Havilland was a British aviation pioneer and aircraft engineer. He started his career in the Crystal Palace School of Engineering. 'The Crystal Palace' was a cast-iron and plate-glass structure, originally built in Hyde Park, London, to house the 14,000 world-wide exhibitors of technology developments for the 1851 Great Exhibition. The facility was moved to Sydenham Hill, London, and reopened in 1854 when it was associated with the School of Art, Science and Engineering. In 1868, the world's first aeronautical exhibition was held in the Crystal Palace.

An engine apprenticeship followed by time as a draughtsman resulted in Geoffrey de Havilland designing an aero engine. The stage was set for the birth of a legend. He would spend his life designing, building and flying aircraft. By 1910 he had designed, built and flown his own successful biplane, barely above the ground for 60 feet (18 metres). In the same year, he joined what would become the Royal Aircraft Factory. In 1912, he established a height record of 10,500 feet (3,200 metres) with his B.E.2 aeroplane. By 1914 he had been 'head-hunted' to become the chief designer of Airco at the Hendon aerodrome. Geoffrey designated all his aircraft with the letters DH followed by a number. During the First World War, he flew with the rank of captain and was a flight commander.

Crash injuries sustained in 1913 prevented him from going overseas, so he was sent to carry out patrols off the east coast of Scotland.

In 1920, he raised the necessary nominal capital of £50,000, equivalent to £2,370,000 in 2018, bought out the relevant aviation assets of Airco and formed the de Havilland Aircraft Company at Stag Lane Aerodrome, Edgware. It was here that he designed the 'Moth' family of aircraft. In 1933, he moved the company to Hatfield Aerodrome in Hertfordshire. The stage was set for the de Havilland Company to maintain the pride of the British aviation industry, gain some research experience and enhance de Havilland's reputation.

Notwithstanding the success of the Supermarine S.6 floatplane in 1931, Britain had no viable competition to the new monoplane airliners being developed in the USA. Determined to make a mark on the world aviation stage, he, with the support of the de Havilland board, proposed a racing aeroplane design specifically for the MacRobertson Air Race, but which would only be built with a commitment of three advance customer orders. The aircraft, the DH 88 Comet, would be subsidised by de Havilland by fifty per cent and would guarantee a minimum speed of 200mph (320kmh).

Orders were received and three DH.88 Comets, designed by Arthur Hagg, were specifically constructed for the race. At de Havilland he invented the differential ailerons used on the Tiger Moth and other de Havilland aircraft. The aileron on the outside of the turn barely moves down, while the aileron on the inside of the turn moves up correspondingly more, to counteract the adverse yaw. The Comet was a stressed-skin cantilever monoplane with enclosed cockpit, retractable undercarriage, landing flaps and two engines with two-pitch variable propellers. The mainly wooden aeroplane had some metal in the high stress areas, such as the engine bearers and undercarriage. The pilots sat in tandem cockpits, inline, one behind the other. The engines were two high-compression race-tuned de Havilland Gypsy Six R engines. Any person familiar with the Mosquito can see its forebear here.

Built in total secrecy, the first Comet flew on 8 September 1934. This was just six weeks before the air race. On 20 October, the three de Havilland Comets started out on the long flight to Melbourne, Australia. One retired, one came fourth and the third, called Grosvenor House after the sponsoring hotel, won the MacRobertson Air Race in a time just over 71 hours. This aircraft was restored to flying condition by the Shuttleworth Collection at Old Warden, England. There were only five Comets built in total. The Comets set many other flight records from 1934 to 1938, such as London to Karachi (October 1934), London to Lisbon

(July 1937), London to Cape Town (November 1937) and London to New Zealand (March 1938).

Geoffrey de Havilland had three sons, Peter, Geoffrey and John. Tragically, two of the sons died as test pilots in de Havilland aircraft. Both were connected to the DH.98 Mosquito. John died in 1943 as a result of a mid-air collision between two Mosquitos. Geoffrey, who had carried out first flights in the Mosquito and Vampire, died in 1946 while testing the DH.108 Swallow. The Swallow was an experimental tailless, swept wing and single vertical stabiliser jet aircraft using the Vampire fuselage as the foundation of the design. All three prototypes had contributed to knowledge about high-speed flight, but at terribly high personal cost.

The de Havilland name had become known throughout the world for preceding aircraft: for the DH series of Moths, such as the DH.82 Tiger Moth and DH.83 Fox Moth; for the DH.89 Dragon Rapide and the previously mentioned DH.88 Comet. With war looming it was a natural progression to look at the success of the Comet and DH.91 Albatross and harness that technology, speed and sleekness to meet Air Force requirements. The British Air Ministry issued specification P.13/36 for a twin-engine medium bomber carrying 3,000lbs (1,400kgs) for 3,000 miles (4,800km) with a maximum speed of 275mph (443kmh) at 15,000ft (4,600m). The Avro Manchester and Handley Page Halifax resulted.

Around any specification there was always talk, conjecture and 'what-ifs' among the various manufacturers' designers. What if the aircraft design went for speed, especially faster than the present fighters, so that the heavy defensive turrets could partially be replaced by fuel and bomb load, leaving the rest for performance enhancement? The possible shortage of duralumin and steel and the Albatross experience with composite ply construction methods and balsawood gave de Havilland the confidence to continue looking at the possibility of using more plentiful wood as a wartime source material.

De Havilland was gaining experience and the DH.95 Flamingo was its first all-metal stressed skin aircraft. However, the Albatross and Flamingo were not providing the solutions to the Air Ministry specifications, that would take a new design and a new DH number. Speed and manoeuvrability were the answer to do away with defensive armament and the third crewman. An October 1938 meeting with the government officials resulted in little interest and this decision would cost them dearly in the timeline to produce the new design, which would be designated DH.98.

The outbreak of war a year later in September 1939 saw de Havilland still advocating an unarmed bomber but that was too radical for the Air Ministry. De Havilland produced drawings and mock-ups at Hatfield to show a defensive gun behind the cockpit to appease the government. Bristol, Blackburn and de Havilland were all proposing a high-speed light reconnaissance bomber. The use of more powerful engines such as the Rolls-Royce Griffon and Napier Sabre was also discussed, allowing defensive armament. Once again, the traditional idea of defensive armament was taking away from the progressive idea of speed and manoeuvrability to survive an enemy attack. Finally, in December 1939, specification B.1/40dh was issued and by January 1940 the DH.98 aircraft was specified with two 1,280hp Merlin 21 ducted radiator engines. By March an order for fifty aeroplanes was placed, the prototype given factory serial number E0234. These would carry a 1,000lb (450kg) bomb with a range of 1,480 miles (2,380 km) at speeds of 397mph (639kmh) at 23,700ft (7,200m).

The Mosquito was unique in that it was mainly constructed of wood; lightness and strength being the two outstanding attributes of the construction. Wood was not a strategically precious material; spruce, birch plywood and Ecuadorian plywood were all readily available. Subcontractors were plentiful as furniture factories, luxury automobile coach builders and cabinet makers could easily adapt their skills to the aviation industry. Battle damage repair was possible in the field without the aircraft being removed from service.

The fuselage, built in mirror-image halves, was made of sheets of balsawood sandwiched between sheets of birch wood. The fuselage bulkheads, seven, were made up of plywood skins separated by spruce blocks. A covering of a fine plain cotton was stretched tightly over the fuselage and then painted with a silver dope. The wing was built as a one-piece structure made up of two main spars, compression ribs stringers and a plywood covering. The wings were covered in the same manner as the fuselage. Engine mounts, undercarriage and the aileron frames and skins were metal. The tailplane was wood, with the control surfaces being metal-framed and covered in fabric. In this way, more Merlin horsepower could be used for performance and not for carrying a heavy fuselage and wing around.

Eight weeks later a fighter, day and night, version was authorised, F.21/40, with cannons and machine guns in the nose and with the addition of Airborne Intercept radar. Shortly thereafter DH.98 was given a name, the Mosquito. By the order of Lord Beaverbrook, Minister of Aircraft

Production, work on the Mosquito was to stop; no materials were made available and production focused on existing types as the Battle of Britain began in July 1940. Air raids took their toll on aircraft production, and the de Havilland factory would not be able to meet its quota of fifty aircraft. In any event, the order was changed to twenty bomber versions and thirty fighter versions. All this work went on in secret at Salisbury Hall, London Colney, Hertfordshire, with non-strategic labour and non-strategic material. The Hall had a long history and at the turn of the twentieth century Lady Randolph Churchill, remarried as Mary Cornwallis West, took up residence there. Her son, Winston Churchill, became a regular visitor.

On 3 November 1940, after two long years or two short years, depending on how you look at it, the prototype E0234 was moved by road the 7 miles (14 kilometres) to the private camouflaged de Havilland facility at Hatfield Aerodrome. It had been bombed one month previously with employee loss of life, destruction of a sheet metal shop and, more importantly, loss of eight months output, by a Junkers Ju 88 bomber; the very target that the fighter version was designed for. The two Merlin 21s, the first variant with Arabic numerals, were available at this time with a two-speed single-stage supercharger driving de Havilland's own three-bladed Hydromatic constant speed controllable-pitch propellers. This engine followed the successful Merlin XX used on the Halifax, Lancaster, Hurricane and Spitfire. The Merlin 21 was capable of +16psi boost with the new 100 octane fuel.

Engine tests were followed by taxi tests by chief test pilot Geoffrey de Havilland on 24 November. On 25 November 1940, the Mosquito went airborne. Further flight testing on E0234, now given the military designation W4050, revealed a rigging adjustment was required to the port wing, that there was a continuing problem of the undercarriage doors opening slightly, and tail buffeting, the latter not cured until February 1941 by aerodynamic fillets on the wing trailing edge and by lengthening the engine nacelles. As it turned out, testing at A&AEE, RAF Boscombe Down, confirmed and exceeded the calculation that the Mosquito would be faster than the Spitfire, by 30mph, 48kmh. The Mosquito had joined the list of aeroplanes requiring Merlin engines. Simultaneously, a new aircraft prototype had just flown in January, the Avro Lancaster. This aircraft would require four.

Certainly, the gauntlet had been thrown down to Rolls-Royce to rise to the production demands of the newly confident Allied air forces. Out of the ashes of Dunkirk and the Battle of Britain, the Phoenix arose to take the battle to the enemy in the guise of the Merlin engine. The Mosquito, during its production lifetime, would require more powerful versions of the Merlin but it all started with the Merlin 21. The aircraft was there, the engine was there, the speed was there.

In April 1941 after demonstration to Lord Beaverbrook, it was decided not only to build the Mosquito in Britain but in Canada, his home country, and Australia, to prevent any major disruption to the supply line. The total number of all types built was 7,781, which included 1,076 in Canada and 212 in Australia, while 3,326 were built at Hatfield, 1,476 at Leavesden, 1,066 at Canley. Others were built at Luton by Percival, and at Portsmouth and Chester by Airspeed. The Americans thought that the wooden Mosquito was not an efficient aeroplane and opted for their own metal Lockheed P-38 Lightning.

By June 1942 the more powerful Merlin 61s, the two-stage two-speed supercharger variant, were fitted with the resulting speed of 428mph (689kmh). It did not stop there, fitted with high altitude two-speed two-stage supercharged Merlin 77 engines a speed of 439mph (707kmh) was attained in December 1943.

Two other airframes, flew seven months later in June 1941: W4051, the photo-reconnaissance version and W4052, the fighter version, with 1,490hp, combat power Merlin 21 engines. The fighter version, apart from the obvious four Browning machine guns in the nose and four Hispano Mk II cannon mounted under the cockpit floor, had a flat bullet-proof windscreen, was painted black, and had the Mk IV Airborne Intercept equipment.

Mosquito W4055 was accepted by RAF 1 Photo-Reconnaissance Unit (PRU) at RAF Benson, Oxfordshire on 8 August 1941. On 17 September, this aircraft and 1 PRU made its operational debut with a secret daylight photo-reconnaissance flight over Brest, La Pallice and Bordeaux, France. Speed and agility were its defence against enemy fighters. That would take some getting used to, especially if the pilot previously flew armed fighters. The Mosquito and the Merlin were now contributing to the war effort. It was how they did it that was different from the few versions available of the Hurricane and Spitfire; the Mosquito had many versions, some sources state forty-one, different series of Merlin engines for example, making up the total of more than 7,700 aeroplanes. These versions were also converted as the need arose and were applied to many varied roles, some very specialised, some not. This does add to the confusion of what designation, F, B, etcetera, and what Mark was doing what, when, and this can only be resolved by following an individual aircraft's history through its lifetime. Every Mosquito, regardless of what it was doing, was powered by two Merlin engines. The types will be discussed in general.

The main types of Mosquitos were photo-reconnaissance, bombers, fighters, night fighters, fighter bombers, trainers, and torpedo bombers. The photo-reconnaissance versions were designated as Mosquito PR Mk I etcetera. The MkIs had the original short engine nacelles and short span tailplanes. The PR Mk I had the Merlin 21 engines giving it a speed of 382mph (615kmh) and for the sake of reconnaissance a range of 2,180 nm (4,040 km). Thirty B Mk IV bombers were converted to PR Mk IV models and flew in 1942.

The five PR Mk VIIIs were converted B Mk IVs fitted with the more powerful 1,565hp Merlin 61 two-stage two-speed supercharged engines. The Merlin 61 was also used in the Spitfire PR Mk XI. This engine incorporated a two-piece cylinder block designed by Rolls-Royce for the Packard Merlin. There was an increase in speed of 50mph over the MkI and a further increase in range to 2,550nm (4,720km) with a ceiling of 38,000ft (12,000m). Continued development led to the PR Mk IX with 1,680hp Merlin 72 engines with droppable external fuel tanks further improving its range. The PR Mk XVI had a pressurised cockpit and used the Merlin 72 1,710hp or Merlin 76 1,655hp engines. The Merlin 73 and 77 had pressurised/cabin pressure blowers.

435 of the PR Mk XVIs were built and had internal bomb bay fuel tanks and droppable external fuel tanks. The apogee of the Merlin engine and Mosquito combination's development were the Mk 32 and Mk 34 range. The PR Mk 34, powered by the 1,690hp Merlin 114 engines, had a massive range of 5,419 miles (8,721km) with internal and external fuel. Complete with the F52 vertical camera and the F24 oblique camera, there was nowhere in the Third Reich that the Merlin could not reach.

The above technical paragraphs show the development of the Merlin horsepower by Rolls-Royce as all the technical innovation went on behind the scenes at the factory in Derby. Yet again, the engine had risen to the occasion and was propelling another aircraft type through the skies to defend Britain and now would be able to take the fight deep into the heart of enemy territory. If the PR Mosquito could reach so far, what about the fighter-bomber version?

The FB Mk VI, which flew with the 1,460hp Merlin 21 engines, had a reinforced wing to carry bombs, 250lb (110kg) or 500lb (230kg), under each wing in streamlined fairings or with the option of mounting eight RP-3 25lb (11kg) or 60lb (27kg) rockets. In addition, provision was made for drop tanks under each wing. The usual fighter armament, of four Browning machine guns and four Hispano cannons, was included. That still left the bomb bay for a similar load as the wings. The bomb bay could instead have overload fuel tanks installed if needed. The versatility of the Mosquito design is encapsulated in this fighter bomber. Nearly 2,300 FB Mk VIs were built, over 30 per cent of total production.

Another fighter-bomber variant was the FB Mk XVIII, known as the 'Tsetse', named after the vicious and dangerous biting fly, that was purpose-built with a Molins '6-pounder' Class M cannon – basically, similar to an anti-tank gun. It had an auto-loader, twenty-five pounds of ammunition and weighed 1,580lb (720kg). The intended target of this gun were heavily armed U-boats on the surface. Additional armour was added to the nose, engine cowlings and under the cockpit floor to protect the aircraft and crew. The Browning machine guns were retained with their tracers as a sighting mechanism for the Molins. The drawback for the crews was that they were a fairly easy target; the gun required a steady approach at a 30-degree angle.

The bomber versions varied as per the photo-reconnaissance versions, namely using the power of the newly available Merlin engines and those aircraft designed for a specific purpose. The Mk IV, by lengthening its bomb bay, could carry four 500lb (230kg) bombs instead of four 250lb (110kg) bombs. A Mk IV was modified, with bulged bomb bay doors and fairings, to carry a 4,000lb (1,810kg) 'Cookie' bomb. Another model was equipped with the 'Highball' anti-shipping weapon, a bouncing bomb designed by Barnes Wallis for the Fleet Air Arm. The B Mk IX, powered by the 70 series of Merlin engines, was designed to allow them to be converted to carry the standard bomb load of 4,000lb (1,810kg) 'Cookies'. The next variant, the B Mk XX, was powered by the US-built Packard 31 Merlin engine and it was the Canadian-built version of the British-built B Mk IV.

The fighter versions similarly improved in performance as the Merlin improved. In the summer of 1942, high flying Junkers JU-86P were proving difficult to intercept with the Spitfire. De Havilland made a decision to develop a high-altitude Mosquito using the Merlin 61 power plants. The main modifications involved the increase in wingspan and the reduction of weight. Test pilot John de Havilland reached 43,000ft (13,000m) in September. Five versions were delivered to RAF Northolt but subsequently the threat disappeared.

The NF, Night Fighter, versions also came in many combinations of Airborne Intercept radar, fuel capacity and firepower. Four B Mk IVs were converted to NF Mk XVs with greater wingspan. In service with RAF 85

Squadron at RAF Hunsdon in 1943, they were reported to have reached altitudes of 44,600ft (13,600m). Night intruders of 100 Group, Bomber Command, had 'Serrate' radar to track radar emissions from the German night fighters and were very successful at finding the enemy aircraft in the night skies.

The Mosquito served extensively in the RAF and the RCAF, RNZAF, RAAF, SAAF and the USAAF. It continued to serve post-war in many countries, both as a military and civilian aircraft. Here are a few operations that were noteworthy above and beyond the more 'everyday' photo-reconnaissance, fighter, bomber, tactical bomber, intruder, maritime strike, pathfinder and fast transport operations. Versatility – nearly fifty different models – was the Mosquito's calling card, performing in the European theatres of war and with the Royal Australian Air Force in the Pacific War.

19 September 1942	First daylight raid on Berlin, Germany.
25 September 1942	Raid on Gestapo Headquarters, Oslo, Norway.
31 December 1942	First 'sky marking' operation with 8 Group Path Finder Force.
30 January 1943	Raid on Berlin broadcasting station, knocking out power during Hermann Goring's speech celebrating the 10th year of the Nazis seizing power
18 February 1944	Raid on Amiens prison freeing French Resistance prisoners
25 March 1944	First U-boat (*U976*) sunk by 'Tsetse' Mosquito
31 October 1944	Attack on Gestapo Headquarters at Aarhus, Denmark
21 March 1945	Attack on Gestapo Headquarters, Copenhagen, Denmark

This last attack was well-planned with large-scale models set up of the building and city to familiarise the crews with the objective. The target was a six-storey U-shaped building. The Underground had provided information that the entire Gestapo staff for Denmark was housed in the Shellhus, or Shell House building. A special force of Mosquitos from RAF 140 Wing was assigned the raid. The Mosquitos took off in three waves of six aircraft, which included two Mosquitos of the RAF film unit. The morning raid ensured that the building would be occupied.

Crossing the North Sea, the Wing made landfall on the coast of Jutland and set course for Copenhagen. The Wing was escorted to and from the target by Mustangs providing top cover for the low-flying Mosquitos. The first wave attacked at rooftop level and one of the aircraft struck a building and crashed. Subsequent photos and reports indicated that the building had been severely damaged by the attack and subsequent fire. More than 150 Gestapo personnel had been killed, while some Danish patriots managed to escape the damaged building.

The Mosquitos were all FB Mk VIs belonging to 487 Squadron, Royal New Zealand Air Force, 464 Squadron, Royal Australian Air Force and 21 Squadron, RAF. The raid showcased the support and co-operation of the Commonwealth Air Forces and their commitment to the de Havilland Mosquito. 487 and 464 Squadrons were Article XV squadrons, formed from the graduates of the British Commonwealth Air Training Plan in Canada, under the operational control of the RAF. All three squadrons had recently moved to France to keep in touch with the advancing Allied armies and were based at B.87/Rosières-en-Santerre.

There is absolutely no doubt that the 7,700+ Merlin-powered Mosquitos contributed to the Allied victory in the Second World War. A Mosquito B Mk IX, *LR503* known as *F for Freddie*, holds the record for the most number of missions flown by an Allied bomber. It flew 213 sorties with RAF 109 and 105 Squadrons; quite an aircraft, quite an engine.

7

Bombers: Avro Lancaster
& Others
1941

What flies slower than two Merlins? Four Merlins attached to an aircraft that could carry a heavy bomb load a long distance. That aircraft is the famous and formidable Avro Lancaster. The ultimate success of this Avro design was made possible by the four powerful Rolls-Royce Merlin engines; the fuselage and power plant combination proved a winning design to take the destructive force of a heavy bomber to the very heart of enemy territory. The Merlin was now daily successfully taking the defensive and offensive Allied aircraft into the skies to carry out their assigned tasks. These were the Merlin's 'finest years'; the fighters, the fighter bombers and the heavy bombers all contributing to final victory.

The Merlin was utilised by many manufacturers as they struggled to design bombers to meet the war effort requirements. Bristol, Fairey, Handley Page, Supermarine and Vickers all created bombers, some more successful than others. They varied from light, medium and heavy bombers to reconnaissance, dive and torpedo bombers. The demand on Rolls-Royce to produce the Merlin would not let up until the war in Europe and the Pacific came to an end. These bombers carried the offensive to the enemy after the successful Battle of Britain defended the island from a planned invasion.

During the First World War, bombs were initially hand-dropped from aircraft on the enemy lines, the seeds of Bomber Command had been sown. During the intervening years – before Second World War aircraft design progress demanded speed and long flights – it was not until the political situation within Germany in the 1930s became apparent that Britain realised the vulnerability of the Air Force and its inadequate bomber capacity. The bomber force at the end of the First World War consisted of the Royal Flying Corps (RFC) de Havilland DH4 and DH9, and the Royal Naval Air Service (RNAS) Handley Page 0/100 and V/1500. All these aircraft were biplanes and, in the case of the de Havilland aircraft, single-engine. In the spring of 1918 the RFC and RNAS were joined together to form the Royal Air Force (RAF).

Another biplane, the twin engine Vickers F.B.27 Vimy, deserves mention. Postwar it was famous for the first non-stop crossing of the Atlantic Ocean from Newfoundland to Ireland. This was closely followed by another long-distance flight, Britain to India. The Vimy was powered by two Rolls-Royce Eagle 360hp (270kW) engines. The inter-war years' bombers were the Handley Page Hyderabad and Vickers Virginia, the last of the mainly wood biplanes. The departure of Sir Hugh Trenchard as Chief of the Air Staff in 1930 changed the RAF focus away from the bomber force. The progress and initiatives of aircraft bomber design were slowed because of this. However, in spite of this change in direction, the inclusion of more metal components in construction signalled the transition to the all-metal aircraft.

The Air Ministry requirement for the bomber to carry up to twenty-four troops hindered the development of the pure bomber. The bomber had a restrictive dual role as a transport aircraft. The Fairey Hendon was the first monoplane bomber to fly but saw very limited service as a light bomber capable of lifting only 1600lb of bombs. The search was now on to find a heavy bomber for the RAF and the following prototypes appeared in response to Air Ministry Specifications. The Armstrong Whitworth A.W.23, the Bristol Type 130 and the Handley Page H.P.51 were the aircraft submitted that still incorporated the requirement to carry troops. A subsequent specification, abolishing the transportation requirement, resulted in a major advance in heavy bomber design. The Handley Page H.P.51 Harrow, first flight 10 October 1936, and the Armstrong Whitworth A.W.38 Whitley, first flight 17 March 1936, were both ordered into production before their first flights took place because although Germany was supposedly restricted to civilian transport aircraft, the gathering strength of the Luftwaffe revealed the military ambitions of the Nazi regime.

Separate specifications had placed the emphasis on speed rather than weight-carrying ability. The results were the Vickers Type 271, later developed as the Wellington, and the Handley Page H.P.52 Hampden. The Wellington actually achieved heavy bomber status with its 4,500lb (2041kg) of bombs carried for 2,000 miles (3219km). The Hampden, meanwhile, was definitely a medium bomber carrying 2,000lb (907kg) of bombs a distance just under 1,900 miles (3,166km).

By 1935 production was underway to replace most of the remaining biplane bombers with metal monoplane aircraft. The Hawker Hind biplane was an exception; it was manufactured to meet the need for a light bomber trainer. The Fairey Battle and Bristol Blenheim joined the bomber fleet as light fast bombers. These decisions coincided with the formation of RAF Bomber Command at Uxbridge on 14 July 1936. Finally, after eighteen years, the RAF had recognised the importance of organising the strike force as a separate entity to the defence force. It was three short years until the bomber force would be called upon to perform its function of carrying destruction to the enemy and returning aircraft and crew safely to home base.

The Air Ministry Specifications had the foresight to measure the distance, albeit initially straight line, between the bases in England and Berlin and include that in their requirements. Also included were the types and shapes of bombs available, as they too were developed. The old machined style became a cast style to improve the aerodynamics and make the behaviour of the bomb more predictable. Initially the utility bomb was 250lb (113kg) but that soon gave way to heavier bombs up to 2,000lb (907kg) and eventually 'bespoke' bombs up to 22,000lb (9,979kg). It would take a special aircraft in the future to carry these large bombs to the target; that aircraft would turn out to be the Avro Lancaster.

Bomber aircraft development continued with two types being considered: the fast light bomber or the heavy load carrying bomber still with its onerous troop-carrying role. The discussion on two versus four engines persisted, the argument for two being that the benefit of four engines was negated by extra weight, drag, maintenance cost, fuel consumption and heavier wing construction. The engines themselves were dividing into two main types, the air-cooled radial or inline liquid-cooled engine. Crew positions were now being specified with the addition of gunners, navigator, wireless operator, bomb aimer and pilot(s). There was no thought to lengthening the takeoff area, so performance had to improve markedly to take the heavier bomb loads off the same airfield.

These were the challenges Roy Chadwick, Avro's chief designer, had to contend with. The story of Avro, Chadwick, and the Lancaster is another chapter in the story of how the Merlin contributed to the war effort. Most of the 7,374 Lancasters that were built used the Rolls-Royce Merlin as their power plant, but the approximately 300 B.II variants used the Bristol Hercules engine. The Lancaster is is just one of nearly twenty Merlin-powered bombers of all types that contributed to the success of Bomber Command. The Handley Page Halifax and Vickers Wellington were two other heavy bombers, but they never achieved the performance of the Lancaster.

Chadwick worked for the A.V. Roe Company, known as Avro, which was founded in 1910. The company was named after the founder Sir Edwin Alliott Verdon-Roe OBE, Hon FRAeS, FIAS, a pioneer aircraft manufacturer and pilot. Although he was born in England, it is interesting to note that he went to Canada in 1891 to train as a surveyor. A downturn in the markets caused him to return to England and seek other opportunities. During his time with the Merchant Navy, he became interested in the flight of the albatrosses he often observed during voyages. A short exposure to the world of aircraft design caused Verdon-Roe to start building models and a subsequent *Daily Mail* winning design in 1907 resulted in his attempt to build a full-size aircraft, based on his winning entry. Despite many setbacks, he became the first Englishman to fly a British-designed and built aircraft, his own Roe 1 biplane, in July 1909. The next year he founded Avro at Brownsfield Mill, Manchester, with his brother Humphrey. A great personal tragedy was the loss of Sir Edwin's two sons, killed while serving with the RAF during the Second World War.

The inter-war years saw the introduction of the Tutor and Anson. The war years saw the Manchester and the Lancaster, which created a number of follow-on types. The York, Lincoln, Lancastrian and Shackleton were all derivatives of the basic Lancaster design. Where did this basic design of the Avro Lancaster come from, who was responsible for the 'outside-the-box' thinking to produce such a successful aircraft? That person was Roy Chadwick CBE, FRAeS. He was following a line of family engineers, which accounted for his interest in aircraft models at age ten. The year was 1903 and the entire world was fascinated with flight as it developed from balloons to gliders and then the thought of powered flight – the sky was now the limit.

In 1918, after studies at the Manchester School of Technology, Chadwick joined the fledgling Avro Company as a draughtsman and personal assistant to Alliott Verdun-Roe. Drafting the first enclosed cockpit monoplane, the Type F, was followed by the 500 series of aircraft leading to the successful 504. Chadwick then went on to design the 504K, which was sold worldwide. During the First World War there were rapidly changing requirements for offensive aircraft and the bomber aircraft became an entity. Chadwick designed the Avro Pike, which was unique in that it had internal stowage for the bombs – shades of the massive internal stowage area of the Lancaster. The Pike also had a gun turret aft of the biplane wings. Continuing his bomber developing theme was the world's biggest single-engine bomber, the Avro Aldershot.

Seaplanes, Arctic exploration, lightweight, ambulance, racer, autogiro, long distance (Avro Avian) aeroplanes all came off the drawing board. The successful Avro Tutor and Cadet, followed by the extremely successful Avro Anson, were all leading to the forerunner of the Lancaster, the Avro Manchester. Chadwick's forethought and brilliance had designed the large enclosed bomb bay for the Manchester. This, in turn, led to the ultimate development of more horsepower, four engines, and the wider wings of the Lancaster.

Just prior to the Second World War, Air Ministry specifications were demanding diversification and there was now a dual requirement, to carry either torpedoes for shipping attacks or the standard bomb load. The torpedoes would necessitate a long compartment if they were to be carried internally. Chadwick had this in mind when he was designing the Type 679 Manchester in response to Air Ministry Specification P.13/36 for a twin-engine bomber. He positioned both wing spars midway in the fuselage centre section to allow an uninterrupted bomb compartment beneath. It certainly was an engineering challenge, but the design decision proved itself invaluable for the future development of the Manchester into the Lancaster. The Manchester Mark I first flew in August 1939 with two Rolls-Royce Vulture 24-cylinder engines, which eventually proved underpowered and unreliable. Less than two years later the Mark III flew with four Rolls-Royce Merlin engines, which, in effect, was the birth of the Lancaster.

Francis K. Mason in his book *The Avro Lancaster* says, 'the Avro design staff, led by Chadwick, had, as early as February 1939, been examining the possibility of introducing a four-engine project, the Type 683, based on the Manchester but with four Rolls-Royce engines.' The Manchester was tested by 207 Squadron, No. 5 Group, at Waddington, by seasoned crews. The cooling problem with the Vulture engine was causing an abnormally high and unacceptable failure rate. The directional problem in the event of an engine failure had been addressed by enlarging the fins and the addition of a central fin, which was later removed. Restrictions were imposed as to performance to lessen the number of engine overheating problems.

The Manchester was relegated to training heavy bomber crew Conversion Units in 1942 until the spares for the Vulture engines dwindled away and the aircraft was grounded. The Manchester production line was now being considered for the Handley Page Halifax and that would have been totally unacceptable to Chadwick and the Avro Company. The commonality of the two Avro aircraft, the Manchester and Lancaster, was pointed out to the Ministry in the hope of getting the go ahead for the Type 683 Lancaster. Avro got the approval for a prototype with the proviso that it had to show its superiority to other types.

A Manchester, BT308, was taken out of the production line and modified with the addition of Merlin X engines, the XX engine not being available at that time. On 9 January 1941 the Manchester III, thus named for security purposes, in effect the Lancaster, went airborne with the central fin in place. By February the Lancaster had been returned to Avro for installation of a wider tailplane, larger fins and rudder and to have the now-unnecessary central fin removed. Further testing at the Airplane & Armament Experimental Establishment (A & AEE) at Boscombe Down revealed that the aircraft was faster than originally designed. This was probably due, in part, to the engine cowlings. The heating and cooling system would require extensive modification due to the variance of temperatures at different crew positions.

Production of the Avro Lancaster was by five companies in addition to the A.V. Roe & Co. factories at Newton Heath, Manchester, Woodford, Cheshire and Yeadon, Yorkshire. The other manufacturers were Austin Motors Limited of Longbridge, Birmingham; Metropolitan-Vickers of Mosley Road and Old Trafford, Manchester; Sir W. G. Armstrong Whitworth Aircraft Limited of Whitley, Coventry, and Bitteswell, Warwickshire; Vickers-Armstrong of Castle Bromwich, Birmingham and Hawarden, Cheshire; and the only overseas manufacturer, Victory Aircraft of Malton, Ontario, Canada, which produced 430 aircraft of the 7,374 Lancaster total.

The Lancaster proved, beyond a doubt, that it was the superior bomber and could carry more bombs farther and with an ease of handling that was important for inexperienced pilots. The tons dropped per aircraft loss are 107 tons for the Lancaster and forty-eight for the Halifax. It had a relatively better accident rate and the casualties were fewer.

The ruggedness of the Lancaster allowed aggressive evasive manoeuvres and allowed the Lancaster to return home often 'on a wing and prayer'. The double waist-high spars held the aircraft together not only in the air but contributed to surviving crash landings.

It is estimated that more than one million men and women worked on producing the aircraft that was the backbone of Bomber Command's heavy bomber fleet. Lancaster: It is a name that evokes memories of single heroic deeds and mass formations of aircraft attacking the enemy and, at the same time. it reminds us all of the great, sadly for many, the ultimate, sacrifice made by personnel in Bomber Command.

There were more Victoria Crosses for bravery awarded to crew members of the Lancaster than any other aircraft. Unlike the fighter pilots who performed their mission alone in their aircraft, and gathered afterwards to regale each other with tales of the events, the bomber crew lived the terror, excitement and satisfaction of a 'job well done' together. The bonding and team spirit of the seven-member crew during wartime was never forgotten and, at the many reunions over the years, those remaining always toasted those crew members that had passed away with the words: 'We will remember them'.

The Lancaster was designed to drop bombs and get home, and this it did very successfully for the duration of the war. This operation was of course not without losses, which varied depending on the weather at the target and at home base, intensity of anti-aircraft (AA) fire, fighter activity, mid-air collisions and the ever-present plain bad luck. The development of radar by Germany, both ground and airborne, contributed greatly to the operational losses. Later in the war radar-guided AA fire and radar-guided ground-controlled night fighters, with their own onboard radar, added to the effectiveness of the German defences. However, by then the tide of war had changed. It was a matter of too little, too late, with dwindling resources. The bombing campaign had taken effect.

In spite of these operational losses the Lancaster flew more than 156,000 sorties and dropped more than 600,000 tons (610,000,000kg) of bombs. The bombs varied in size from incendiary 4lb (1.8kg) bombs, sea mines, 250lb (9,113kg) bombs to the mighty Grand Slam at 22,000lb (9,979kg). Mention should be made of the specialty bombs, such as the rotating skipping bombs used in the famous 'Dambusters Raids'. Initially operations were flown during the day, quickly switching to night raids due to the heavy losses. Towards the end of hostilities, with diminishing fighter resistance, daylight raids resumed with long-range fighter escort. The raids were conducted at all altitudes from 'on the deck' penetration raids to mid- and high-level saturation bombing. The number of aircraft varied from Squadron level to the all-out massive assault of the three '1,000-bomber' raids.

The bombing campaign of the first half of the war had been carried on with a collection of aircraft that would eventually be mostly replaced by the Lancaster. Bomber aircraft such as the Armstrong Whitworth Whitley, Bristol Blenheim, Fairey Battle, Handley Page Hampden and the Short Stirling carried on the bombing campaign until development overtook them and they were assigned to training and other duties. The Whitley was assigned to Coastal Command, the Blenheim to night fighter duties, the Battle to air gunnery schools and target tug duties, the Hampden as a torpedo bomber with Coastal Command, the Stirling to mine-laying, troop carrier and glider tug, and the Wellington took part in the 1,000-bomber raids in 1942 before relinquishing major duties to the Lancaster.

Air Chief Marshal Sir Arthur Harris, KCB, OBE, AFC, assumed the post of Commander-in-Chief Bomber Command in February 1942, coinciding with the introduction of the Lancaster to Bomber Command. The end of December 1941 saw the delivery of three Avro Lancasters to 44 (Rhodesia) Squadron at Waddington with instructions to have eight crews fully operational by the end of January 1942. Wartime did not allow for the luxury of long conversion courses; the Lancaster was needed yesterday. It flew its first operational flight against Germany on 3 March 1942. This was a 'gardening' flight to lay mines against the German Navy and shipping.

Harris was soon to change the direction of the bomber force. The present mining operations, 'gardening missions' to support the Navy, were displaced to second priority in favour of mass strategic bombing that was designed to destroy the industrial structure and demoralise the general population. Opposition to this bombing policy was centred on the theory that bombing targets should be restricted to military installations only. Many great minds have debated this issue and it appears in many books but this book will not be one of them. Instead, the focus will be on the history of what raids the Lancaster took part in.

Harris saw first-hand the destruction caused by the Luftwaffe as the bombs rained down on London. He was a witness to the challenges of fire suppression and to the morale of the general population as they went about their daily lives amid the chaos and aftermath of a bombing raid. 7 September 1940 saw the start of nearly two months of continuous bombing of London, the Blitzkrieg. It is said that Harris stated, while standing on the Air Ministry roof during one of these raids, 'They have sown the wind, and so they shall reap the whirlwind', a reference to Hosea in the Bible. There was no doubt from now on about the future bombing policy of the soon to be nicknamed 'Bomber' Harris.

On 10 March 1942, two weeks after his assuming command, the first operation over Germany by Lancasters occurred, a raid on Essen. The Harris mandate had begun and also the ensuing controversy. The mandate would continue, the controversy would not deflect him. The background on the mandate was the Butt Report of 1941. Bomber Command had no way of verifying crew claims of targets bombed and damage done, and it was decided to mount cameras triggered by the bomb release which would record the event.

Sir Winston Churchill, the Prime Minister, made his famous 'The Few' speech on 20 August 1940 which is remembered for his remarks about the gallantry of the fighter pilots but included in that speech were the words 'On no part of the Royal Air Force does the weight of the war fall more heavily than on the daylight bombers who will play an invaluable part in the case of invasion and whose unflinching zeal it has been necessary in the meanwhile on numerous occasions to restrain.'

The heavy bombers such as the Avro Manchester, Handley Page Halifax and Short Stirling joined Bomber Command in 1941 and, although an improvement, the Command still lacked striking power. The Manchester and Stirling in particular were a disappointment as a heavy bomber due to lack of power preventing them carrying heavier bomb loads. Something new was needed to improve the productivity of the bombing campaign. The something turned out to consist of four parts to form the formidable bomber force, they were the leader, Harris; the new aircraft, the Lancaster; the new technology, GEE; and the new target identification technique using flares, the forerunner of the Pathfinders.

The open weight of the Lancaster for fuel and bombs was approximately 21,000lb (9,525kg). The longer the distance, the more fuel that was needed, with less available capacity for bombs. The direct route was preferred but – due to the searchlight, radar and gun defences – was not always available. The whole raid was a compromise, with fuel being the priority. However, in spite of careful planning, a stronger than normal westerly wind on the return journey could have the crews making an emergency landing away from base at an East coast runway or, worse, ditching in the Channel. Taking a few of the more common targets, using Lincoln as a departure station as an example, the distances were as follows with the number of Main Force attacks on the target.

Raids	Target	Distances Round trip miles/kms	Flying time Straight line*
28	Essen	684/1102	6.0
24	Berlin	1168/1880	8.5
22	Cologne	712/1146	6.0
18	Duisburg	664/1070	6.0
18	Stuttgart	1048/1686	7.5
17	Hamburg	872/1400	7.0
16	Hannover	866/1394	7.0
13	Mannheim	934/1504	7.0
12	Bremen	776/1248	6.5
11	Nuremberg	1130/1820	8.0
2	Königsberg	1720/2770	10.0

*After consulting some log books, it seems the actual flight time was 2 to 2.5 hours longer on average than the straight-line distance due diversion around known defences, evading night fighters and en-route tactics.

Harris stated, on many occasion, words to the effect that the Avro Lancaster was the greatest single factor in winning the war because of its performance. It could lift a 22,000lb (9,979kg) bomb, it could absorb punishment both from enemy fire and evasive manoeuvres, and most importantly could bring the crews home against all odds. He referred to the aircraft as 'that shining sword in our hands'. The double waist-high wing spars were an important part of its robustness. Its long bomb bay was its crowning glory and enabled the Lancaster to carry ordnance of all shapes and sizes, sometimes with necessary modification to the bomb doors. A Lancaster variant, the B.VI Pathfinder aircraft, was fitted with the Merlin 85/87, which had two-stage superchargers, giving much improved high-altitude performance. However, the nine aircraft converted were not a success and were withdrawn in 1944. That particular Merlin engine proved troublesome, rough running and difficult to synchronise, for the crews and squadron maintenance.

The following statistics are for relative information only and not to be accepted as totally accurate; they were gathered from many sources, which often disagreed but, taken in context, will give an idea of the quantities involved in events that happened more than seventy years ago. In many cases they have been rounded off to the nearest 100. Wartime, by its very nature, would not lend itself to accurate bookkeeping, the emphasis being on getting the job done, update the records later. The statistics themselves were often not adequately described or defined, leading to apparent discrepancies. Some indicative statistics follow:

1 Ton equals 1,016 kilograms

1942	35,338 aircraft despatched	1,450 missing	45,561 tons dropped
1944	166,844 aircraft despatched	2,770 missing	525,518 tons dropped
1939–1945	657,674 tons dropped on Germany		
1939–1945	955,044 total tons dropped		
1942	35,637 tons on industrial towns		
1944	184,688 tons on industrial towns		
1942	12 tons on oil targets		
1944	48,043 tons on oil targets		

All types highest number of sorties despatched,	day	1189 16 November 1944
Sorties despatched on one target,	day	1,107 Dortmund 12 March 1945
Highest tonnage dropped on one target,	day	4,851 Dortmund 12 March 1945
Sorties despatched on one target,	night	1,047 Cologne 31 March 1942
Heavy bomber despatched on one target,	night	970 Duisburg 14 October 1944
Highest tonnage dropped on one target,	night	4,547 Duisburg 14 October 1944
Highest number of aircraft missing,	night	96 Nuremberg 30 March 1944

Air combat claims destroyed	1,191,	probable 310,	damaged 897
Aircraft type destroyed	JU88 – 333,	Bf110 – 198,	ME262 – 5

Bomber Command crew lost or missing in action   55,573

Lancaster (Halifax for comparison in brackets)

Despatched 148,403	Attacked 135,445	(Despatched 73,312, Attacked 66,456)
1942 tons of bombs dropped	11,367	(7,274)
1944 tons of bombs dropped	361,004	(146,113)
Total tons of bombs dropped	608,612	(224,207)
Total Lancasters built	7,377	(6,176)
Missing aircraft	3,345	(1,833)

The monthly average of Bomber Command sorties started with 1,900 in 1939 and built to a crescendo of 16,800 by 1945. Similarly, the monthly total tons of bombs and sea mines dropped grew from 1,600 in 1939 to an enormous 46,200 in 1945. These figures were not achieved without great personal loss and sacrifice. The loss rate started at 4.1 per cent in 1942, it was 3.7 per cent in 1943, 1.7 per cent in 1944 and 0.9 per cent in 1945. The introduction of Radio Counter Measures (RCM) in 1944 had an initial dramatic effect on losses until Germany introduced its own RCM. However, by then the tide of aerial combat had turned with the Allied Forces on Continental Europe in June and the subsequent occupation of France and Belgium taking away the early warning radar capability. These are the average losses for the year, but some catastrophic losses still occurred, such as the nearly 60 per cent loss rate of the daylight Augsburg raid on 17 April 1942.

What were the results for all this effort? It is generally credited with creating restrictions in the movement of raw material to production facilities and disrupting industrial planning. An inordinate number of persons were required for defence purposes rather than production purposes. The psychological effect of night-time bombing by the Lancaster, in addition to the casualties, is very difficult to quantify. Suffice to say it was not a positive experience and therefore it must have had an effect on the population and defence forces.

The Lancaster is an inanimate machine, without its supporting crew it is nothing. Unless it was maintained, repaired, armed, refuelled, and the crew was fed, housed and kept healthy, the whole effort would be for nought. Then it was up to the bravery and skill of the Bomber Command crews who produced the destruction statistics that contributed to the Allies winning the other battles in the air, on and below the sea, and on the ground, that ultimately won the Second World War.

The Lancaster, although arguably the most successful of the heavy bombers, as mentioned previously, was not the only bomber in the RAF.

Frederick Handley Page, who would become known as the 'father of the heavy bomber', started his aviation career studying electrical engineering. He joined the Royal Aeronautical Society, where he befriended Jose Weiss who was experimenting with wing design in gliders. Similar to Verdun-Roe, he learned to fly and attempted several aircraft designs around the same time. He established Handley Page Limited in 1909. During the initial stages of the First World War, Handley Page designed the large twin-engine bomber, the 0/100, which subsequently was developed into the 0/400 and the four-engine 0/1500 bomber. All these aircraft used the Rolls-Royce Eagle engine, thus establishing a good business relationship for the future.

During the inter-war period, Handley Page devoted its research to wing aerodynamics and invented the 'Handley Page slat', which was a narrow opening running the length of the wing leading edge to improve airflow at high angles of attack. The twin-engine medium bomber HP.52 Hampden led to the HP.56 proposal of the twin-engine Rolls-Royce Vulture aircraft, on which the Air Ministry changed the specification to a four-engine heavy bomber, the HP.57 Halifax. This aircraft would initially be powered by the Rolls-Royce Merlin X engine, which would later be replaced by the more powerful Rolls-Royce Merlin XX.

The Halifax entered service in November 1940, fourteen months ahead of the Lancaster, and had its debut with No. 35 RAF Squadron on operations to bomb the harbour at Le Havre in occupied France. During this period, it bore the brunt of the daylight bombing casualties that eventually led to the late 1941 decision to undertake night sorties only. By 1942, No. 35 RAF Squadron had been selected to form the core of the new force, the Pathfinder group, which eventually became No. 8 Group.

Initially, the general opinion was that the Halifax was limited in its potential performance by the Rolls-Royce Merlin X engine. This was improved by the use of the Bristol Hercules engine and the Rolls-Royce Merlin XX engine in succeeding Marks of the Halifax. It was also limited by the capacity of the bomb bay being unable to carry the 4,000 'Cookie' blast bomb. However, it did play an important part in the 1,000-bomber raid on Cologne, Germany, on the night of the 30/31 May 1942. It is not far off the mark to estimate that there were 3,000 or more Merlin engines on that raid. Each Merlin would have to be assembled from Rolls-Royce and subcontracted parts, tested, transported to the squadron, installed, air tested and maintained for this great effort. Air Chief Marshal Harris, after receiving congratulations from Frederick Handley Page, replied that the success of the raid 'was very largely due to your support in giving us such a powerful weapon to wield. Between us we will make a job of it.'

The Halifax continued to be manufactured through 1943 when it equipped all of No. 4 Group and would be operated by all fourteen squadrons of No. 6 Group Royal Canadian Air Force. Bomber Command at one time had seventy-six squadrons equipped with the Halifax. By August 1944 the full offensive was on, as indicated by the first daylight operation against the oil refinery at Homberg in the Ruhr Valley.

The Merlin also powered the Halifax in its many submarine warfare, reconnaissance and meteorological sorties with Coastal Command. The Halifax, in addition, was called upon as a glider tug, electronic warfare aircraft, and to work with the Special Operations Executive. The Merlin, along with the Bristol Hercules, formed a partnership to allow the Handley Page Halifax to contribute in a meaningful way to the Allied victory.

As mentioned previously the Merlin was used in quite a few other bomber aircraft, not always as the exclusive engine for the type. One of the first bombers to use the Merlin in its infancy was the Armstrong Whitworth Whitley. It was one of three front-line medium bombers in RAF service at the outbreak of the Second World War. The other two were the Vickers Wellington and the Handley Page Hampden. The initial Whitley versions were powered by the Armstrong Siddeley Tiger engines until replaced by the Rolls-Royce Merlin IV and X engines. The Merlin was accredited with greatly improving the Whitley performance. The Whitley was designed for night bombing and No. 4 Group, equipped with the Whitley, was a fully trained night bomber force at the start of the war.

The Vickers Wellington was continually produced during the war, being produced in a greater quantity than any other British bomber, more than 11,400 of them. Vickers studied the age-old question of air-cooled versus liquid-cooled engine. In the end both types of engines were used, the air-cooled Bristol Pegasus, Hercules and Pratt & Whitney Twin Wasp engines and the liquid-cooled Rolls-Royce Merlin X and later the Merlin R6SM engine. The Wellington took part in the first bombing raid of the war on 4 September 1939, bombing shipping at Brunsbüttel.

The Avro Lincoln long-range heavy bomber, developed by Chadwick at the end of the war, was based on improving the Lancaster and was powered by the 66, 68A, 85, 102 and 300 two-stage supercharged Merlin engine. Too late for active war service, it did continue serving in the RAF for nearly fifteen years. The Fairey Barracuda was a British carrier-based torpedo and dive bomber, the first all-metal aircraft for the Royal Navy's Fleet Air Arm. It was powered by the Merlin 30 and 32 engines. The Barracuda is famous for scoring fourteen direct hits on the German battleship *Tirpitz* on 3 April 1944. Operation Tungsten was flown off the carriers HMS *Victorious* and *Furious*.

Other Fairey aircraft using the Merlin were the Battle light bomber and the Fulmar light dive bomber. The Battle used the Merlin I, II and III Merlin engines. It took part in the Battle of France as part of the Advanced Air Striking Force. Two Victoria Crosses were awarded to a battle crew during the 12 May 1940 attack on Albert Canal road bridges despite overwhelming defensive fire. Experiencing continuing heavy losses as its performance could not match the current aircraft, the Battle was withdrawn to a training role. The 'Hurribomber' was the Hawker Hurricane IIB and IIC, modified to carry bombs on hard points under each wing. It was powered by the Merlin XX engine.

The contribution of Bomber Command to the Allied victory is without question; destruction, disruption and intimidation rained down upon the enemy. The Merlin powered the aircraft that allowed the RAF and Allied Air Forces to bring these forces against the Axis powers. However, the technical genius of the Merlin and its various aircraft would all have been for nought had it not been for the brave crews that carried out these operations, the ground crew, women and men, who maintained the Merlin-powered aircraft and the entire support staff, both military and civilian, who through their supreme effort, kept the various organisations effective. 'A chain is as strong as its weakest link.'

8

Fighter Escort:
North American Mustang
1942

Perhaps influenced by the previous 1,000-aircraft RAF raids, the United States Army Air Force (USAAF, 1941–1947) believed, initially, in the power of numbers for protection as it scheduled mass formations of bombers during daylight raids to attack Germany. The weakness in this idea was that the gun turrets on the bombers could not move fast enough to track a high-speed attacking fighter. The higher-flying Luftwaffe fighters would be able to pick their targets from above and would quickly inflict damage as they passed through, in, and out of, the large bomber formation. Numbers just meant more targets of opportunity, not efficient protection.

Two raids in particular were an indication that something had to change. During the raids on Schweinfurt, in August 1943, and Regensburg, in October 1943, 120 bombers were lost, which was too high a price for the supposed pin-point accuracy of the daylight raids. The RAF had already abandoned daylight bombing raids because of the high loss rate and changed, like the Luftwaffe, to night raids. The bomber formation would need protection during daylight or the USAAF would have to change to night operations. The approach of Bomber Command was to continue the bombing campaign twenty-four hours a day so that the USAAF change to night flying was not considered an option. The mistaken belief was still predominant among the American generals that, accordng to Douhet's theory on war, the 'bomber will always get through' and they also believed that the heavy bomber could defend itself as it penetrated 'the thin wall' of the German fighter defences.

What was required was a fighter that matched or exceeded the performance of the present enemy fighters to protect the bomber formation by flying 'top cover'. The problem was that the fighters did not have the range of the bombers and initially during the campaign the bombers were left exposed to enemy fighter attack at their most vulnerable time, closest to the heavily defended target as they manoeuvered for their run-in and began their journey home. That special USAAF fighter would turn out to be the North American P-51. Named the Mustang, the word is defined in dictionaries as 'the small, hardy, half-wild horse of America'. It was not the Allison-powered Mustang, it was the P-51B Mustang powered by the Rolls-Royce Merlin engine that made it extra special. The Merlin made all the difference and turned the Mustang into arguably the best American (or even Allied?) single-engine piston fighter of the Second World War.

Where did this Mustang come from? Who was North American Aviation and was this a one-off or did they manufacture any other planes of note? Who were the people behind the design and the big question is how did an American aircraft end up with the most sought-after British engine? How could Rolls-Royce keep up with the demand or was there another player in the game? The answers to these questions provide the story of the Mustang and its rise to World War fame and lasting legacy. In this case, the aeroplane did not make the engine, the engine made the aeroplane, the Merlin in its finest American hour!

The founder of North American Aviation was Clement Keys. He has been described by Harry Bruno, himself a pioneer aviation public relations figure and tireless aviation promoter, as 'the father of commercial aviation in America'. Keys, who was born in Canada, was a financier specialising in aviation-related companies, an aviation entrepreneur. In the 1920s he was involved with such companies as Curtiss-Wright, Trans World Airlines and the China National Aviation Corporation. In December 1928 he established North American Aviation (NAA) as a holding company, which bought and sold shares in many different aviation businesses. The Air Mail Act of 1934 would dramatically change NAA.

The political decision to deliver mail via the Army Air Corps had failed and the government returned air mail delivery to the commercial airlines, with some restrictions. It restored competitive bidding, prevented old companies that previously held contracts from obtaining new ones and dissolved holding companies, such as NAA, that brought airlines and aircraft manufacturers together. NAA was forced to change its focus if it was to remain in the aircraft industry. It became an aircraft manufacturer under Kindelberger.

James 'Dutch' Kindelberger became NAA's first President in 1934 and brought J 'Lee' Atwood with him. They had both previously worked for Douglas Aircraft Company since 1930, working on the DC-1 and DC-2 transport aircraft. Kindelberger's strategy was to manufacture small training aircraft and stay away from head-on competition with the larger aircraft companies such as Boeing and Douglas. The NAA NA-16, low-wing monoplane with tandem seating, built in 1935, was the first training aircraft and was the forerunner of 17,000 trainer aircraft that NAA would build. The later model T-6 Texan World War II advanced trainer was known as the Harvard by the RAF and RCAF. It was the most famous and widely used Allied trainer of the war. It is still in demand in 2018, both as a private aircraft and as an airshow demonstration aircraft. The noise of the Pratt & Whitney 600hp (450kW) radial engine power plant is impressive, especially so if in a formation.

Kindelberger moved the NAA factory from Dundalk, Maryland, to El Segundo (Los Angeles), California, to take advantage of the better all-year-round flying weather. By 1936 the factory housed 250 employees. NAA later opened factories in Dallas, Texas, and Kansas City to produce in excess of 43,000 aircraft, more than any other American manufacturer from 1938 through 1945. Another member of the team was Edgar 'Ed' Schmued, an aircraft preliminary design engineer. Born in Hornbach, Germany, in 1899, he was self-taught in engineering and started work as an apprentice in a small engine factory. In 1925 he left Germany and joined General Aviation, the air branch of the General Motors Corporation (GMC), in Brazil. In 1931 he emigrated to the US to work for another GMC subsidiary, Fokker Aircraft Corporation of America. Throughout the war years NAA was also, a little known and surprising fact, a division of GMC.

Although successfully established in the manufacturing of training aircraft, NAA had the facilities and expertise to manufacture other types of aircraft – and the US Government was aware of its capabilities. Simultaneous with the development of the Mustang, NAA developed a very successful twin-engine medium bomber, the B-25 Mitchell. Nearly 10,000 Mitchells were built using two Wright Twin Cyclone 14-cylinder 1,700hp (1,267kW) engines.

The British Purchasing Commission (BPC) was established in New York, USA, to arrange the production and purchase of weaponry from North American manufacturers. To bypass the Neutrality Acts that were in place, there was a 'Cash and Carry' arrangement using Britain's gold reserves. As we haver seen, there was an aeroplane shortage in Britain during the initial years of the conflict, through lack of foresight and sluggish reaction to the Axis build-up of wartime resources. By the end of 1940 aircraft deliveries were 300 per month and expected to grow in numbers by 1941. This economic and political opportunity stimulated design, and NAA was among the companies to take advantage of this. Designs were given a name by the British Air Ministry to replace the American numbering system. To maintain neutrality, all aircraft had to be shipped through Canada either by crate on a ship for smaller aircraft or flown trans-Atlantic by RAF Ferry Command, the weaponry could not go directly to Britain.

In March 1941, a Lend-Lease agreement was signed by the USA to supply the Allied Forces with food, oil and materiel. The materiel included ships, aeroplanes and other weaponry. This effectively ended USA neutrality as she realised the Allies were struggling in Europe and elsewhere and an Axis victory would certainly affect the US. At that time no one could foresee the attack nine months later on Pearl Harbor in December 1941.

Previously, after testing at A&AEE Boscombe Down, the Lockheed Lightning P-38 had been turned down as unsuitable. Conditions were now favourable for a new aircraft to be developed in America. Bureaucracy usually ruled the meetings of the manufacturers and the BPC. The initial idea was for NAA to upgrade the Curtiss P-40 and improve its performance, always a demand in wartime conditions, and build it under licence. The BPC believed that Kindelberger, Atwood and Schmued and their team were quite capable of doing just that. NAA was already building the Harvard trainer for the RAF so Britain already had experience with the NAA product. The BPC asked NAA to obtain copies of the blueprints and technical data of the P-40 and the abortive XP-46 to examine them. After paying for this information, it was found by Edward Horkey, NAA's chief aerodynamicist, that the 1936 information was already obsolete. The question was whether to build an obsolete aeroplane or take their experience and build their own state-of-the-art fighter. The team decided that NAA could build a better product and the NA73 (Mustang) would result from that decision; thus the RAF had an American aeroplane designed for them.

The Mustang, first and foremost, was a British-driven NAA-engineered aeroplane, into which the USAAC had no direct input. The aeroplane was procured on a direct contract prior to the introduction of the Lend-Lease programme. The Air Corps had no connection with the specifications, flight tests or acceptance of the initial production models. In fact, there was resentment against the aircraft and its British connection until, eventually, its outstanding performance could not be ignored when it was compared to the existing USAAC aeroplanes.

The Curtiss P-40 used the Allison V-1710 liquid-cooled V12 engine. The Allison Engine Company was founded by James Allison, a race car team owner, who had made his money as a partner in an acetylene company. The company was located in Speedway, Indiana, which was located close to what is now the Indianapolis Motor Speedway. Allison established an engineering company to look after his fleet of race cars. The company established itself with world-wide recognition for work done on the Liberty aeroplane engine and a steel-backed bronze bearing. Norman Gillman was the chief engineer who, using the experience gained on the Liberty engine, designed the liquid-cooled V12 Allison V-1710.

After initially building engines for the US Navy, by 1937 the V-1710 had become the engine of choice for the new United States Army Air Corps (USAAC, 1926–1941) fighters, the Lockheed P-38 Lightning, Bell P-39 Airacobra and Curtiss P-40 Warhawk. A V-1710-C6 completed the USAAC type test at 1,000hp (750kW) in April 1937, reputedly the first engine of any type to do so. Hence, the engine for the NA73 would be no exception in the fighter group. By 1941 the Allison Engineering Company, a division of General Motors, had three Indianapolis plants with a work force of 12,000. General Motors developed a single production line, and a basic power section modular design that could be adapted to various propeller gearing systems and superchargers.

The rear of the engine was the accessory section and could be easily adapted to different superchargers, magnetos, and fuel and oil pumps. The engine also featured the ability to rotate in either direction by the addition of an idler gear system. The initial engines were specified to have a single-stage supercharger and thus limited their performance at altitudes above 15,000ft (4,600m). This limited the V-1710 aircraft to theatres of war other than Europe, where the need was for high-altitude capability. An attempt was made to improve high-altitude performance using turbo-superchargers; using exhaust gases to turn the turbo-supercharger, not gearing from the engine crankshaft. Successful with the radial engine, the turbo-supercharger was problematic with the liquid-cooled V-1710.

The BPC took Kindelberger at his word that NAA could design a new and better fighter than the P-40, using the Allison V-1710 engine, and in April 1940 signed an agreement for the Model NA-73X prototype to be completed in 120 days! Such were the demands and vagaries of wartime. Schmued followed the conventional practices of the era with the addition of several features, such as the laminar flow wing developed by NAA and the National Advisory Committee for Aeronautics (NACA). These laminar aerofoils generated very low drag at high speeds. However, you never get anything for nothing. The NA-73 lacked lift at low speeds and required large flaps to keep landing speeds low. Another newly discovered feature was the 'Meredith Effect' of heated air leaving the radiator with a slight amount of thrust.

The fuselage comprised mathematically-computed conic sections that gave smooth, low drag surfaces. The main landing gear retracted in to the wing and was covered with doors. The steerable tailwheel, unlike the Spitfire, retracted in to the fuselage and was covered by two doors. All in all, the NA-73 was what could be termed an aerodynamic 'slick machine'. A streamlined single duct under the rear fuselage provided air for the cooling and lubricating heat exchangers. The internal fuel capacity was greater than the Spitfire, giving an increased range of operation. In spite of initial engine deficiencies, the fuselage held great promise for development and was far superior to any of its contemporaries.

At the outset NAA designed the assembly lines to be suitable for mass production, as previously invented by Henry Ford for his automobiles. The sub-assemblies would move through various stations for installation of the respective components and systems. A full-scale mock-up with associated templates was used to facilitate the build-up of systems. Nearly 3,000 drawings were prepared for the prototype. The co-ordination of the design team with the shop people enabled a prototype to be constructed in four months, somthing which, on previous occasions, had taken three years. Not only had the assembly of NA73 to take place in an orderly fashion but also the ease of disassembly had to be considered. Each production aircraft would have to be built, test flown, disassembled and crated for shipping to England. On 7 October the Allison V-1710-39 engine arrived, and its installation was completed on 11 October. The engine had a single-stage, single-speed supercharger supplying air to the Bendix Stromberg two-barrel carburettor. The V-1710-39 was rated for takeoff at 1120hp at 3,000rpm and 44.5 inches manifold pressure. An initial problem was the tendency of the engine to overheat when run at high power settings, the radiator scoop shutters would have to be fully open during this time.

Rolled out in September 1940, the NA-73 first flew on 26 October 1940. This was five days before the generally accepted end of the Battle of Britain when the war would turn offensive instead of defensive. Subsequently the prototype crashed on its fifth test flight in California. No aerodynamic or mechanical fault caused the accident, the selected fuel tank had run dry. NAA had to make the immediate decision to continue with the Mustang or abandon it in favour of the Curtiss-Wright

P-40; thankfully, sane heads prevailed, the Mustang development would continue. The NA-73 was 25mph (40kmh) faster than the Curtiss P-40 powered with the same engine. The RAF started to place orders for the NA-73 and NA-83, with fishtail ejector exhausts, and designated them the North American Mustang Mk I. By April 1941 the first British production aircraft came off the assembly line.

620 Mustang Mk Is with bullet-proof windscreen and self-sealing fuel tanks, 320 NA73 and 300 NA83s were built at the Inglewood factory in California and supplied to the RAF. A year after its first flight in California, the P-51s had been shipped to England via the Panama Canal and the first demonstration flight occurred at the end of October 1941. Meeting its required flight statistics, the Allison-powered Mustang Mk I was prepared for pilot training and to join the list of Allied fighters on strength.

The first Mustang Mk Is entered service with the RAF early in 1942 and were assigned to Army Co-operation Command for tactical reconnaissance and ground attack duties due to their poor high-altitude performance. These aircraft were paid for by the British Government. Subsequently, 93 aircraft were the first Mustang P-51s Lend-Leased to the RAF and were designated Mk Ia. The last of the pre-Merlin Mustangs, the P-51A-NA, were designated the Mk II. Fifty were Lend-Leased to the RAF. All in all, there were 713 Allison-engine Mustang aircraft used by the RAF for low-level operations. These were the initial aircraft from NAA backed by the industrial might of the USA, the forerunners of more than 15,000 P-51s from the NAA factories.

The Mustang almost passed into history because of the lack of a more suitable power plant and some political and military manoeuvring by the USAAC procurement channels preventing ordering of the aircraft. Before the Mustang, in 1933, exercise reports had been published in the USAAC decrying the use of pursuit aircraft in favour of 'high speed and otherwise high performing bombardment aircraft'. Colonel H. H. (Hap) Arnold, Chief of Staff, USAAC, finished a report by saying: 'No known agency can frustrate the accomplishments of a bombardment mission.' Seven years later, in 1940, Arnold, now Chief of USAAC, originated the most sweeping change to fighter development in American Air Force history. Reality, experience and a finally open mind had caught up with the limitations of bombardment, perhaps influenced by the actual British experience before America entered the Second World War.

Captain Claire Chennault had challenged the validity of the invincible bomber mentality of that 1933 exercise by devising a defence system based on ground observers and radio communication. Incurring the wrath of his superiors, because he had gone against Douhet's theory of accumulating vast amounts of bomber aircraft, the matter of fighter, pursuit and escort aircraft was pushed further away from active Air Corps thought. Besides, no fighter could ever fly as far as a bomber aircraft. How times and thoughts would change. A counter to every weapon can be devised given time, finances and commitment and the bomber aircraft were no exception; for them it would be the fighter.

Post the Great Depression, the American aircraft industry had languished somewhat and then suddenly, in 1938, Britain and France were shopping in the free world for aircraft to supplement the short supply of the excellent fighters they already had as the threat of the European war loomed. President Roosevelt, in September 1938, held a conference in which he authorised a large increase in aircraft production. The USAAC had to contend with the unexpected USA government purchase of aircraft and the competition for aircraft with foreign buyers. When France capitulated, the 4,000-aircraft order was taken over by Britain. The money received from these foreign orders was approved to be used for development of aircraft for the USAAC. NAA used this money to develop the Mustang. The American manufacturers benefitted from the reports sent back from the war zone on how their equipment was performing.

What event would change the Mustang from a good aircraft to a great fighter-escort aircraft? This event is of the same magnitude as when Stanley Hooker was put in charge of all Merlin supercharging, after which the engine and Rolls- Royce would never be the same. A few months after the Mustang was in service, in April 1942, Ronnie Harker, the Rolls-Royce service-liaison pilot, was invited to go to the Air Fighting Development Unit at RAF Duxford by a colleague, Wing Commander Ian Campbell-Orde, to try out the new American fighter. This is Ronnie Harker's account of that flight as recounted in his book, *Rolls-Royce from the Wings*, published in 1976, as quoted in *The Magic of a Name: The Rolls-Royce Story, The First Forty Years* by Peter Pugh and published by in 2000:

The General took me to the aeroplane and showed me all around it, explaining as he did some of the history of how the specifications had originated. As I flew the Mustang, I felt that it had a number of desirable features which the current fighters lacked. I was particularly impressed by its large fuel capacity of 269 gallons on internal tanks. This was three times as much as the Spitfire. I also liked the six .5 heavy machine-guns

Above left: C.S. Rolls. (Courtesy of Rolls-Royce Heritage Trust)

Above right: F.H. (Henry) Royce. (Courtesy of Rolls-Royce Heritage Trust)

Rolls-Royce
Eagle, the first
aero engine.
(Courtesy of
Rolls-Royce
Heritage Trust)

Rolls-Royce R
engine, 1931.
(Courtesy of
Rolls-Royce
Heritage Trust)

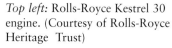

Top left: Rolls-Royce Kestrel 30 engine. (Courtesy of Rolls-Royce Heritage Trust)

Top right: Supermarine S6 at Calshot. (Courtesy of Rolls-Royce Heritage Trust)

Above left: Hawker Hart with Merlin engine, 1938. (Courtesy of Rolls-Royce Heritage Trust)

Above right: Merlin upper crankcase. (Courtesy of Rolls-Royce Heritage Trust)

Left: C.S. Rolls Cross-Channel commemorative statue, Dover. (Courtesy of Jennifer Stevens)

Packard Merlin 29. (Courtesy of K. C. (Colm) Egan)

BCMC Avro Lancaster Packard
Merlin 224. (Courtesy of David Birrell)

BCMC Avro Lancaster Packard
Merlin 224. (Courtesy of David Birrell)

Hawker Hurricane Packard Merlin 29. (Courtesy of Richard de Boer, Calgary Mosquito Society)

Hawker Hurricane Mark IIC. (Courtesy of Keith Burton)

Supermarine Spitfire Mark IX. (Courtesy of Geoffrey Pickard)

Supermarine
Spitfire Mark
VB. (Courtesy
of Keith
Burton)

Supermarine
Spitfire Mark
TR9. (Courtesy
of Malcolm
Nason)

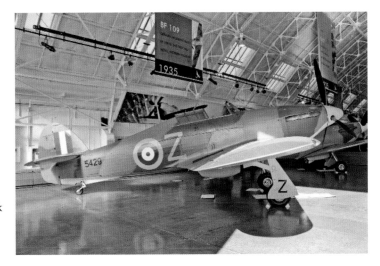

Hawker
Hurricane Mark
IIA. (Courtesy
of Malcolm
Nason)

Above left: De Havilland Mosquito Merlin 113/114. (Courtesy of Richard de Boer, Calgary Mosquito Society)

Above right: Rolls-Royce Battle of Britain memorial window. (Courtesy of Rolls-Royce Heritage Trust)

De Havilland Mosquito T3. (Courtesy of John Mounce)

De Havilland Mosquito FB26. (Courtesy of John Mounce)

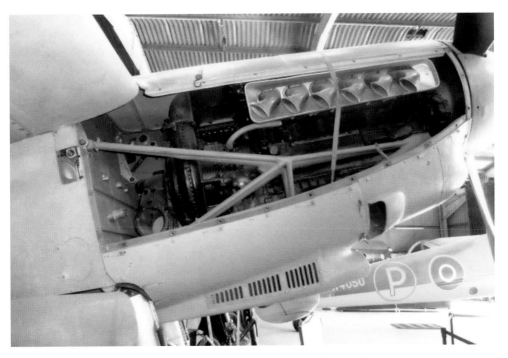

De Havilland Mosquito B35 Merlin 114. (Courtesy of John Allen)

Avro Lancaster Merlins at war. (Courtesy of Canadian Museum of Flight)

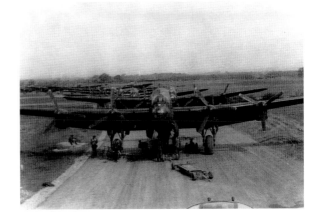

Avro Lancasters 419 Squadron, Middleton, St George. (Courtesy of Canadian Museum of Flight)

BCMC Avro
Lancaster Flight
Engineer's panel.
(Courtesy of
Wayne Ralph)

Canadian
Warplane Heritage
Museum's
Avro Lancaster
C-GVRA.
(Courtesy of Keith
Burton)

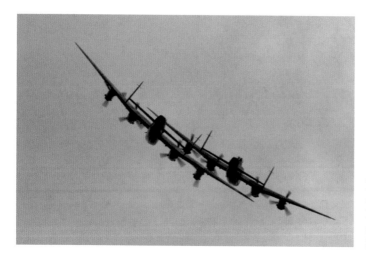

Eight Merlins
at Portrush,
Northern Ireland.
(Courtesy of Bill
Powderly)

BCMC Avro Lancaster Packard
Merlin 224. (Courtesy of David
Birrell)

North American P-51D Mustang,
NL335J, 1983. (Courtesy of John
Kimberley)

North American P-51D Mustang.
(Courtesy of Keith Burton)

Voodoo, the record-setting P-51D
Mustang. (Courtesy of Harry
Measures)

Steve Hinton, record setter. (Courtesy of Harry Measures)

Rolls-Royce Shadow factory, Hillington, Scotland. (Courtesy of Rolls-Royce Heritage Trust)

Above left: Trans Canada Airlines Avro Lancastrian Merlin. (©Air Canada Archives)

Above right: Canadian Warplane Heritage Museum's Avro Lancaster on a 2010 airshow brochure. (Courtesy of Eddie O'Brien)

2015 Abbotsford, Canada airshow. (Courtesy of Jerry Vernon)

Canadian Warplane Heritage Museum's Avro Lancaster Packard Merlin 224.
(Courtesy of CWHM)

Bomber Command Museum of Canada's Avro Lancaster Packard Merlin 224. (Courtesy of Douglas Bowman, BCMC)

Bomber Command Museum of Canada's Avro Lancaster night run-up. (Courtesy of Douglas Bowman, BCMC)

Above left: The Avro Lancaster ND654 engine being recovered in Belgium. (Courtesy of Huyghe-Decuypere Group)

Above right: Martin Rouse's Mosquito Merlin engine. (Courtesy of Martin Rouse)

Martin Rouse's Mosquito Merlin engine pistons. (Courtesy of Martin Rouse)

Merlin Propeller shaft, at Vintage V-12's. (Courtesy of Atushi 'Fred' Fujimori)

Merlin crankshaft, at Vintage V-12's. (Courtesy of Atushi 'Fred' Fujimori)

Above: Merlin valve train at Vintage V-12's. (Courtesy of Atushi 'Fred' Fujimori)

Right: Juan Dorador repairs a Merlin wheel case at Vintage V-12's. (Courtesy of Atushi 'Fred' Fujimori)

Restored Merlin, at Vintage V-12's. (Courtesy of Atushi 'Fred' Fujimori)

Merlin on a test stand at Vintage V-12's. (Courtesy of Atushi 'Fred' Fujimori)

mounted close inboard in the wings, the light and effective aileron control which gave a high rate of roll and perhaps most important of all, its low drag, which gave it a very noticeable increase in top speed over both British and German contemporary fighters.

With the low full throttle height of the Allison engine its overall performance was adequate for Army operations and reconnaissance duties at low altitude. However, one saw immediately the possibility of the Mustang as an air-superiority and long-range penetration fighter – if only it could be fitted with our latest two-speed, two-stage supercharged Merlin. If this was successful, it could be the answer to both the Me 109 and the Focke-Wulf 190, thus providing certain qualities that the Spitfire lacked. I discussed this proposal with the General after my flight, and we agreed to put it up as a serious proposition.

On returning to Hucknall, I asked Witold Challier, our Polish performance expert, to estimate what the Mustang would do if when fitted with a Merlin 61. He reported that there would be a greatly improved rate of climb and an increase of some 40mph in top speed at 25,000ft and above. This estimate, together with the fact that her tank capacity would give her longer range meant that the Mustang, when fitted with the Merlin, would be superior to any other fighter at that time.

There is a memory, quotation or interpretation discrepancy regarding the engine recommended for the Mustang conversion. In his book *The Engines were Rolls-Royce*, published in 1979, Harker states:

> I was very impressed with the aeroplane, which seemed to me to be a natural for the two-speed, two-stage Merlin 66, which was just coming into service on the Spitfire IX. I felt that the Merlin 66 would greatly improve the speed of the Mustang, as its full throttle height would be very much greater than with the Allison engine and the 30 or so miles per hour higher indicated speed maintained at altitude would mean an increase in true speed of perhaps 50 miles per hour! I discussed this with Campbell-Orde and Ted Smith after the flight and they concurred, so we agreed to try and promote the proposal of re-engining the Mustang with the Merlin.

The Merlin engine numbers could, at times, be quite confusing and not so easy to recall. The Merlin 65, according to Gruenhagen in his book *Mustang: the Story of the P-51 Fighter*, was a development engine designed specifically for testing the P-51. The engine differed from the Merlin 61 in that it had a pressure carburettor and a new two-stage blower assembly, a bigger impeller but turning at slower speed. However, the important point is that Harker recognised the potential of the Mustang with the Merlin engine and succeeded in having the engine changed, which certainly was another example of 'The Engine that Won the Second World War'.

Harker had difficulties persuading the powers-that-be but eventually won over Ernest Hives, then Rolls-Royce Director and Works Manager, for his idea of the engine replacement. The drawback was that the Air Ministry wanted every Merlin 61, two-speed two-stage intercooled supercharger, installed in Spitfire IXs to combat the high flying Focke-Wulf 190. By May 1942 a Mustang I was sent to the Rolls-Royce Experimental Flight Test facility at Hucknall. In June Hives scheduled a meeting and Major Thomas Hitchcock, Assistant Air Attaché at the American Embassy in London, was present. Later Hitchcock said in a briefing '... The whole story of the English fighter planes is more a story of engines than it is of the planes themselves. When you talk about engines, you practically get down to the Rolls engine – that is the Rolls Merlin engine.' And later in a subsequent briefing '... by putting the Merlin 61 into the Mustang. They believe that will be the best fighting plane for the next year or two; and their preliminary tests indicate they are right.'

Challier, the performance expert, indicated that a Merlin 61 engine Mustang would have a service ceiling of 40,000ft and full throttle height of 25,000ft. These figures astounded the British and were not believed in the USA, so much so that they sent over Dr. Edward Warner, Vice Chairman of the Civil Aeronautics Board, to verify the statistics. Warner verified Challier's performance graphs. The RAF and USAAC now had a potential aircraft capable of improved performance over its peers. By October 1942 Rolls-Royce had flown their first prototype Mustang X conversion, with the medium-low altitude Merlin 65 engine, to be followed on 30 November by NAA's prototype, XP-78, later renamed XP-51B. The NAA conversion would never have happened if it were not for the strong urging and lobbying by Major Hitchcock to the American Government to proceed with the conversion. The Merlin 65 series engine was utilised in all the prototypes as it was identical to the Merlin 66 powering the Spitfire Mk IX, allowing for a closer comparison.

On 13 October 1942 Captain R. T. Shepherd, the Rolls-Royce Company test pilot, flew Mustang AL975-G on its first flight since conversion to the Merlin engine. The flight was disappointing for a few reasons, although not totally unexpected. The speeds were lower than anticipated due to pressure building up in the engine cowlings causing it to become loose, the main radiator ducting was bigger than required and was reduced for further tests, and the test altitude was restricted due to the supercharger being operated in MS mode because of unmodified fuel pumps. Daily flight tests with subsequent modifications saw the speed attain 422mph in FS gear, still below calculated value.

The Merlin 61 could be installed into the Mustang with no alteration to the cowling using a Consus (Convergent Suspension) engine mounting. Rolls-Royce, however, was worried about the supply of the high-altitude Merlin 61 destined for the Spitfire and initially offered the Merlin XX instead. Challier's graphs showed the Merlin XX Service Ceiling as 36,300ft and the Merlin 61 as 40,100ft. The Merlin 61 was called the Packard V-1650-3 in the US. Various engines, such as the Merlin 65, 66, 70 and 71, and propeller combinations were flown in the Mustang at Hucknall, the Rolls-Royce Experimental Flight Test facility, during the conversion flight test programme.

The Mustang was not without its teething troubles. The Aeroplane and Armament Experimental Establishment (A&AEE) had flown the second conversion, AM208, and commented on the need for additional trim to maintain directional stability with a change of power and speed and that the aircraft tended to sideslip easily during manoeuvres. These characteristics would remain with the Mustang until the taller fin model P-51H. A constant problem was the undercarriage doors opening, becoming unlatched, at the higher speeds in a dive. This problem, remarkably, would take more than a year to get fixed and unfortunately was the cause of several unexplained accidents, with aircraft breaking up in flight. A warning device was installed belatedly to show that the door was not closed.

What happened on the other side of the Atlantic with NAA and the Packard Merlins? Simultaneous development of the P-51 Mustang caused continuous exchange of construction information and testing data between NAA in Inglewood, California, and Rolls-Royce in Derby. The Allison V-1710 engine had been a reliable, trustworthy workhorse of an engine and now NAA was working with a thoroughbred racehorse engine. The V-1650-3 high performance power plant gasped for air below normal barometric pressure, it would lurch in its mounts during major power changes or when the supercharger went to the high-blower position, it would search, with a galloping rhythm for its new stable power setting. During engine deceleration when landing there would be a shower of burning carbon from the exhaust stacks flowing by the cockpit and through it all would be the noise, the unforgettable noise, of the V-12 Merlin!

The two-speed impeller supercharger was controlled by an aneroid sensor, which would switch it with a 'lurch' between low- and high-speed settings. The high-speed setting would rob the engine of 200hp to drive it. The compound function throttle would automatically drive the engine controls for throttle valves, manifold pressure regulators and the magneto spark advance. Six weeks after the British Mustang had its first flight, the American Mustang took to the air – but not before curing the ground run engine overheating problems caused by clogging of the radiator by deposits from the cooling lines.

The first test flight on 30 November 1942 was aborted after 45 minutes due to the old problem, engine overheating. The problem was diagnosed as electrolytic action between the glycol and the dissimilar metals of brass and copper fittings with the aluminium radiator causing scale to loosen in the cooling system. A new radiator and scoop redesign with exit shutters controlled by thermostatic bulbs cured most of the problem. The result, a P-51 designated XP-51B, that had a 10,000ft (3048m) greater ceiling, 50mph (81kmh) greater top speed and a climb time to 30,000ft (9144m) of 12.5 minutes.

The first production versions were crated in August 1943 and arrived in Britain in September. The crating was changed to a cocoon wrap to protect the aircraft against the elements during shipping as it provided a quicker assembly. The first five aircraft were stationed at Boxted, near the historic town of Colchester, Essex, with the 354th Fighter Group on 11 November 1943 as part of the 9th Air Force and later the 8th Air Force. The group's primary role would become as escort to the Allied bombers on their missions over the Continent and eventually all the way to Germany. The 354th pioneered the fighter tactics that would be required to escort large bomber formations over Germany.

The pilots not only had the culture shock of adapting from the tricycle gear Bell P-39 Airacobra they had flown at home to navigating the tailwheel NAA P-51 around the irksomely non-geometrical English countryside. Colonel Don Blakeslee left no doubt in his briefings about his take on the terrain, British weather, the Axis fighter pilots, and air discipline. He demanded that his men 'never break away from a German making a head-on attack.' His pilots were to make the enemy pilot break, shoot him down, or run into him, never, never veer. His pilots were to demonstrate to the enemy that they owned the sky.

They became known as the 'Pioneer Mustang Group' and would achieve the distinction of having the most number of air victories, 701, of any Fighter Group in the European theatre of Operations, as well as the only Congressional Medal of Honour recipient, Colonel James Howard. Howard was escorting the 401st Bomber Group when he became separated from his Fighter Group and single-handedly attacked thirty enemy fighters until he ran out of ammunition. He shot down six becoming an 'ace-in-a-day', damaged others, and somehow got away in the confusion. Merlin and Man at their finest.

The new aircraft required that the ground crews and pilots be combat-ready by 1 December. Twenty-four Merlin-powered P-51Bs, led by Colonel Blakeslee, departed Boxted at 1429hrs on 1 December 1943 for their first mission. It was a fighter sweep of Saint-Omer in the Pas de Calais region of France. Some flak damage ensued, with no enemy aircraft sightings, and the formation returned safely at 1549hrs.

The first escort mission was four days later on 5 December 1943 with the 1st Air Division. Two wings of Boeing B-17 Flying Fortresses and two wings of escort P-51B Mustangs took part in a raid on Poix, southwest of Amie, and the P-51B Mustangs would be relieved by Republic P-47 Thunderbolts with external tanks.

For a short time in 1943 it was proposed that a special Boeing B-17F 'Destroyer Escort' would escort the bomber formations and fourteen were built. The YB-40 had special heavy armour plating, Bendix chin turret, seven pairs of .50 calibre machine guns and a Martin turret at the aft end of the top of the fuselage. It carried a massive amount of ammunition for the guns. The idea was scrapped as the YB-40's performance caused the bombers to slow down approaching and leaving the target area, which was not an enviable outcome.

Harker's farsighted idea, drop tanks and the Merlin engine, turned the Mustang P-51B/C, and subsequently the D model, into the finest escort fighter and all-round fighter of the Second World War. An example early on of the capability of the P-51B was the bombing raid on Keil, Bremen and Hamburg. This was the first mission that the Forty-Five P-51Bs flew with 75-gallon (341-litre) drop tanks and they flew the 500-mile (805km) route as escort fighters for the bombers. Two hours and forty minutes after takeoff, the 1,462 bomber aircraft, the largest bombing force to date, hit Kiel. They still had to fly home. Some of these escort flights could last up to seven hours in a cramped cockpit, not improved by the addition of winter flying clothing and life preservers. The British did improve the cockpit by the installation of the Robert Maxwell balloon canopy for better room and vision, but it did have one drawback, no frame to block the sun. The lighter-weight aircraft with only four guns was faster and crisper to fly and many felt that it was the best variant of P-51. By the end of 1943, 361 escort sorties would be flown by the Merlin P-51Bs in the European theatre. The Merlin floodgates had opened with the arrival of the P-51B/C and the Mustang III to Britain, Packard had taken the Rolls-Royce design and married it to one of the finest fighter airframes of the time. The Merlin-powered P-51D was not far behind.

Most of the P-51B/Cs were assigned to the 8th, fighter-escort, and the 9th, strike force, Air Forces in England, with fewer going to the 12th and 15th Air Forces in Italy. They were the prime Mustang variants until the arrival of the P-51D, starting in March 1944. The second fighter group assigned initially to the 9th Air Force and then switched to the 8th was the 357th Group for fighter escort duties. It flew its first mission on 11 February 1944. This was very quickly followed by the 363rd Group on 23 February 1944. The fighter escort component was building rapidly.

Simultaneously the 12th and 15th Air Forces were assembling in Italy and the Fighter Groups at Madna, Lesina, Ramitelli and San Severo. The Merlin-powered Mustangs were replacing the Supermarine Spitfires and the 31st Fighter Group flew its first escort mission on a bombing raid to the Ploesti oil refineries in Romania on 21 April 1944. Following the 52nd Fighter Group to join the 15th Air Force was the 332 Fighter Group, a unique Group of black pilots that trained at Tuskegee, Alabama. The Army was a segregated service in those years. Their top scorer was Lea Archer with five air kills, one kill was later reallocated to another airman to prevent Archer being recognised as an ace.

During 1944 the Fighter Groups established themselves in England at such quaint sounding places as Debden, King's Cliffe, Steeple Morden, Fowlmere, Bodney, Leiston, East Wrotham, Botisham, Wattisham and Rivenhall; locations where the echoes of Merlin engines can still be heard by those with acutely sensitive hearing. The P-51B/Cs were also active in the China-Burma-India theatre from September 1943, flying 100,000 combat hours until November 1945.

The first RAF Mustang IIIs arrived at No. 65 Squadron in December 1943, followed by No. 19 Squadron in March 1944, both at Gravesend, Kent. These squadrons were quickly followed by twenty-eight other squadrons including Polish, Canadian, Free French and a Coastal Command squadron. By February 1944 the Mustang III was an integral part of the RAF medium and USAAF heavy bomber escort services. NAA delivered 274 P-51Bs and 626 P-51Cs, Mustang IIIs, to the RAF with a perceived weakness, in that they had only four guns instead of six and these were susceptible to jamming. The anti-jamming modification to the P-51Ds were later retrofitted to the Mustang IIIs.

The 281 P-51Ds supplied under the Lend-Lease agreement of 1944 were designated Mustang IV for the RAF. A further 594 P-51Ks were also designated Mustang IVs. During the invasion of Normandy, the RAF Mustangs were used in the fighter-bomber role with the 2nd Tactical Air Force. By the end of 1944, the aircraft had reverted to Fighter Command and some were used in the interceptor role shooting down the V-1 flying bombs.

And of course there was also the USAAF. By the end of the Second World War the USAAF had developed and grown into virtually an independent service. Post-war in 1947 it became the United States Air Force. The USAAF theory of strategic bombing was initially formulated at the Air Corps Tactical School, Maxwell Field, Alabama. The theory supported the sustained precision bombing of railways, harbours, cities, workers' housing and industrial areas by heavily armed long-range bombers. This was to break the enemy's will to resist, to destroy his military, political, economic, and social life.

Initially the daylight operations of unescorted bombers were not as successful as planned, the attrition rate from enemy fighters was too high. The arrival of the Merlin-engine Mustang with its internal fuel and external jettisonable fuel tanks now meant that the bombers could be protected on their longer and longer raids into the heart of Germany. No longer were vulnerable bombers relying on numbers for safety, they were now shepherded by squadrons of Mustangs to the target and back. Some of the longer target distances were Berlin, Dresden, Nuremberg and Munich, taking up to seven hours with the necessary doglegs to avoid the enemy defences and concentration of fighters. It was a long, solitary, uncomfortable, tense escort for the single-engine Mustang pilot who was responsible for the protection of the aerial convoy. At least the pilot did not have to worry about his aircraft, the Merlin engine and Mustang airframe provided him with a very capable fighter to match the enemy.

The bomber aircraft were divided into three categories, the Very-Heavy B-29, the Heavy B-17 and B-24, the Medium B-25 and B-26 Bombardment Group. The B-29 Boeing Superfortress was primarily used in the Pacific Theatre, best suited to its long-range target operations. A natural development from the Boeing B-17 Flying Fortress of the European Theatre, the high-altitude pressurised B-29 dropped the atomic bombs on the Japanese cities of Hiroshima and Nagasaki. With the very long distances involved, it had been up to the ground forces to move forward with the general advance and capture islands at great loss to provide operating bases for the P-51s escorting the B-29s.

The B-17 Flying Fortress was half of the heavy bomber force in Europe. The other half was the Consolidated B-24 Liberator, the American bomber type with the most number of aircraft built, just over 19,000. The Flying Fortress was second in number, just over 12,000. After an ignominious start, when the prototype crashed owing to the control locks being left engaged and so could not complete the competition, the USAAC realised the importance of the aircraft and orders began to mount, slowly at first, at the end of the 1930s. Of the many variants, the G model (August 1943) was most widely built, followed by the F model (May 1942). The G model had increased the number of guns from seven to thirteen. The RAF operated Fortress 1s as bombers in 1941 with little success. The C model was shortly assigned to Coastal Command for bombing of U-boats, eventually sinking eleven of them.

The USAAF began building up its forces in Europe after the Pearl Harbor attack on 7 December 1941. The Eighth Air Force arrived in England in May 1942 and established the 97th Bomb Group. The first mission was in August with B-17Es, escorted by two separate groups of Spitfires, one to and the other from the target, against the railroad marshalling yards at Rouen, France. It was a moderate success with little enemy opposition. At that stage the question whether the bombers could penetrate deeply into Germany without fighter escort all the way was unanswered.

In 1944 the B-24 comprised half of the USAAF bomber force and featured with the Fifteenth Air Force out of Italy. The RAF also used the Liberator in Coastal Command on similar duties with the B-17. It had an aerodynamic 'Davis' clean wing and was the first American bomber to have a tricycle undercarriage, unlike the tailwheel of the B-17. It generally was not as robust a design as the B-17 and was more susceptible to battle damage; its better performance came at a price.

Unescorted raids continued for another year, with the result that a raid on Regensburg and Schweinfurt resulted in a horrifying loss rate of 16 per cent of the 376 bombers deployed. This was not the true extent of the mission disaster, another 20 per cent were permanently lost to operations due battle damage. It was believed at the time that the technology did not exist for an escort aircraft to fly fast enough and have sufficient range and performance to exceed, or at least meet, the best of the enemy's fighters. The Luftwaffe tried to escort their bombers over Britain with the Messerschmitt Me-110 Destroyer. It failed one of the three aforementioned requirements, it did not have the performance to meet the Hurricanes and Spitfires in aerial combat. Schweinfurt, with a

26 per cent loss rate in October 1943, screamed out the need for an escort fighter that had a greater range than the Republic P-47 Thunderbolts and Lockheed P-38 Lightnings.

On 20 February 1944 the Eighth Air Force dispatched more than 1,000 bombers to Europe in 'Operation Argument'. They were accompanied by seventeen Fighter Groups (835 fighter aircraft) and sixteen RAF squadrons for shorter escort coverage. The targets were twelve different major assembly and component factories. On this date, the loss rate was 2.4 per cent. The Merlin engine in the new long-range Mustang played a major role in this success. Postwar statistics indicate that between 225 and 275 Luftwaffe fighters were lost during 'Argument'.

Other fighters may have equalled the NAA P-51 in one or more combat roles, but no other fighter has equalled it in all its many roles. It turned out to be a most versatile, rugged, but yet aerodynamically refined aircraft as a Dive Bomber, Close Air Support, Photo Reconnaissance, Fighter Bomber and Long-Range Fighter Escort. It served with thirty RAF squadrons, three RCAF squadrons, as the Mustang III (P-51B/C) and Mustang IV (P-51D). Not only did it prove itself in the Second World War but continued service in the Korean War in the early 1950s and in many foreign skirmishes.

Air Marshal Sir William Sholto Douglas, Commander-in-Chief RAF Fighter Command, said: 'The Mustang after it had been built with our Merlin engines, which were built under licence in the United States, became one of the finest and most versatile of all the fighters that were produced during the Second World War.'

PART 4

THE MERLIN DEVELOPS
1933–45
The Fame Grows

9

Variants, Shadow Factories, Packard

The exigencies of war and the Allies' determination had a tremendous influence on the development of their respective societies. Driven by the desire to win the conflict there were great improvements in industry, scientific research and the medical field. It was not all wine and roses though. This was war. The death and destruction that daily assailed the Allied population tested the resolve of the people. There were few families not directly affected by war and its consequences. For communities in the eastern half of Britain, there was the daytime aerial noise of the USAAF bombers' Pratt & Whitney and Wright engines and at night-time the noise of the RAF bombers' Rolls-Royce Merlin and Bristol Hercules engines. There were the attacking Luftwaffe's engines and its bombing campaign with the accompanying fighters. To round off the cacophony there was the sound of the defending fighters, mostly Merlin engines.

In spite of the daily disruption of their lives, rationing, lack of sleep, the men and women of Rolls-Royce and subsidiary component manufacturers were able to continue to develop the Merlin engine over the six long war years. They took it from its humble beginnings of 740hp (552kW) to 1,670hp (1,245kW), as used in the Spitfire LF Mk XVI, an increase in power of 123 per cent. The Merlin was the right engine for the right aircraft, at the right time.

The search for the perfect Merlin engine continued, engines could always be better, work never stopped. Merlin engines were designed for many different applications and performances: high altitude, low altitude, medium altitude and even sea-level for Coastal Command. The struggle towards a perfection that could never be realised continued unabated for Packard, Rolls-Royce, and their subcontractors and component suppliers. This is why the Merlin was so successful an engine for so many years. Royce had created a foundation that was good enough to allow for such progress, there was no need to rip it all up and start again. The following tables (*see* pages 185 to 187 and pages 191 to 198) illustrate some basic information to show the parallel development of the Merlin engine by Rolls-Royce in England and Packard Motor Car Company in the USA. The tables should be used as a general understanding of how the Merlin developed during the war and not as a complete list. It was difficult to verify statistics as reliable source were sometimes in conflict.

The following is an alphabetical list of the manufacturers of the aircraft types in the table that used the Merlin engine: Armstrong Whitworth Whitley, Avro Lancaster, Avro Lincoln, Avro Tudor, Avro York, Boulton Paul Defiant, Bristol Beaufighter, Canadair North Star, Curtiss Warhawk, de Havilland Hornet, de Havilland Mosquito, de Havilland Sea Hornet, Fairey Barracuda, Fairey Battle, Fairey Fulmar, Handley Page Halifax, Hawker Hurricane, Hawker Sea Hurricane, North American Mustang, Supermarine Seafire, Supermarine Spitfire, Vickers Wellington, Westland Welkin. These aircraft cover a wide range of types and manufacturers built to fulfil a wartime need and contribute to the ultimate victory, each deserving its place in aviation history. They should not be forgotten for their contribution, big or small, and the contribution of the people involved who created and maintained them. Some of these aircraft, such as the Beaufighter, Tudor, Warhawk, Wellington and Whitley, had other engines installed as well as the Merlin. The following is a brief description of these aircraft.

Armstrong Whitworth Whitley
The Whitley was a twin-engine medium bomber, which first flew on 17 March 1936. Designed as a night bomber, it first dropped propaganda leaflets over Germany. In March 1940 it conducted the first raid on German soil, bombing the Hornum seaplane base. Shortly thereafter, on 11/12 June 1940, it bombed the cities of Turin and Genoa a few hours after Italy had declared war. Soon obsolete, it was relegated to paratrooper resupply duties and served with Coastal Command. In the spring of 1942, British Overseas Airways Corporation (BOAC) used the Whitleys as freighters on such runs as Gibraltar to Malta and Scotland to Sweden.

Avro Lincoln

The Lincoln was a four-engine heavy bomber which first flew on 9 June 1944. Developed from the Avro Lancaster, it was destined for the Pacific War, which ended before the Lincoln saw World War II service. It did see service with the RAF in Malaysia and Kenya. During the Cold War a Lincoln was shot down by Soviet MIG 15s on a flight to Berlin. It also served with the Australian and Argentinian Air Forces. The Lincoln III later became the Avro Shackleton and served with the RAF and South African Air Force.

Avro Tudor

The Tudor was the first pressurised British airliner. It was based on the Avro Lincoln and first flew on 14 June 1945. The tailwheel configuration did not help against its competition and after a chequered history, modifications and accidents, BOAC refused to accept it as part of their Trans-Atlantic fleet and cancelled its order for Tudors in 1947, instead taking delivery of Canadair North Stars, which they renamed C-4 Argonauts.

Avro York

The York was a transport aircraft based on the Avro Lancaster and first flew on 5 July 1942. The York had military and civilian roles with various operators between 1943 and 1964. In civilian service, British South American Airways and BOAC were the main users of the type. In military service, the York was prevalent during the high-profile air-supply missions during the Berlin Blockade, from June 1948 to September 1949. They flew 58,000 supply runs carrying more than 1,000,000 tons of supplies. It was also used as a VIP head of state transport, ferrying such passengers as Winston Churchill, the British Prime Minister, and the French General, Charles de Gaulle.

Boulton Paul Defiant

The Defiant was a British interceptor aircraft which first flew on 11 August 1937. It was a turret fighter, no forward firing guns, and after mounting losses reverted to a night fighter role where it had more success using Airborne Intercept (AI) radar. By 1942 it was target towing and providing electronic countermeasures, 'Moonshine', against the German Freya radar. A formation of eight Defiants could simulate a formation of 100 aircraft, so diverting the enemy away from the Allied bombing formation. On 11 May 1945, Martin Baker, now recognised as the world-famous ejector seat manufacturer, used the Defiant for their ejector seat trials.

Bristol Beaufighter

The Beaufighter, based on the Bristol Beaufort, was designed as a multi-role aircraft which first flew on 17 July 1939. Similar to the Defiant, it was later equipped with AI radar for night interceptions. The aircraft was also operated by Coastal Command, Fleet Air Arm and the RAAF in its Southwest Pacific campaign. The USAAF operated the Beaufighter in the Mediterranean area. A unique claim to fame for the aircraft occurred on 12 June 1942 when a single RAF 236 Squadron Beaufighter flew low level to Paris and insolently dropped the Tricolore, the French flag, on the Arc de Triomphe before strafing the Gestapo Headquarters in the Place de la Concorde.

Curtiss Warhawk

The Warhawk was an American fighter and ground-attack aircraft, which first flew on 14 October 1938. The aircraft was widely used by the Allies, using the names Tomahawk and Kittyhawk for different models, and was operational world-wide in the different theatres of war. The Warhawk was the aircraft used by the legendary Chennault's 'Flying Tigers', an American Volunteer Group flying with the Chinese Air Force. RCAF Kittyhawks shot down Japanese balloon bombs over British Columbia, Canada. The balloons were designed to cause wildfires throughout North America after floating across the Pacific Ocean. Aircraft were also supplied to the Soviet Union by the Alaskan-Siberian ferry route.

De Havilland Hornet/Sea Hornet

The Hornet further developed the wooden construction methods of the Mosquito and first flew on 19 April 1944. It was primarily for the long-range war in the Pacific and resembled a scaled-down single-seat Mosquito. The Hornet had contra-rotating propellers to counter the effect of torque. It did not see action in the Second World War but in 1951 was involved in the Malayan Emergency. The Sea Hornet was the carrier-based version.

Fairey Barracuda

The Barracuda was the first metal carrier-borne torpedo and dive bomber for the Royal Navy's Fleet Air Arm, which first flew on 7 December 1940. In 1943 they deployed to the Mediterranean and supported the Salerno Landings. On 3 April 1944 Barracudas from HMS *Victorious* and *Furious* disabled the German battleship *Tirpitz*.

Fairey Battle

The Battle was a single-engine light bomber, which first flew on 10 March 1936. It was the first operational aircraft to be powered by a Merlin engine entering service before the Hurricane. During the 'phoney war' Battles were deployed to France as part of the British Advanced Air Striking Force. On 20 September 1939 a Battle shot down a Messerschmitt Bf 109, credited as the first aerial victory of the war.

Fairey Fulmar

The Fulmar was a carrier-borne fighter aircraft, which first flew on 4 January 1940. The prototype had previously flown on 13 January 1937. It had long range and was used in the Mediterranean Theatre and for reconnaissance of enemy fleet movements. Fulmars destroyed 112 enemy aircraft, which made it the highest scoring Fleet Air Arm fighter of the war.

Hawker Sea Hurricane

The Sea Hurricane was the naval version of the Hurricane. It came in two distinct versions, the carrier version with catapult spools and arrester hooks, or the catapult armed merchantman (CAM) ships version, known as 'Hurricats'. Recovery was by ditching close to a ship and being picked up or, fortunately, recovering on nearby land. The catapult Hurricanes started development in January 1941 and the first aircraft was sent to the FAA in March 1941 for trials. The first true Sea Hurricane – with the ability to take off and land on an aircraft carrier – began with a conversion of a Mk IIC (BD787) in May 1942.

Supermarine Seafire

The Seafire was the naval version of the Spitfire and first flew on 7 January 1942. The initial versions were classified as marginal when employed in the naval environment, suffering from handling problems with the narrow undercarriage track, low speed handling difficulties, short range and lacking strength for hard landings. Later versions improved these problems and the Seafire began to replace the Sea Hurricanes. Their first combat use was in 'Operation Torch' in November 1942, the Allied invasion of North Africa, followed by the invasion of Sicily and Italy during the summer of 1943.

Vickers Wellington

The Wellington twin engine long-range medium bomber first flew on 15 June 1936. A unique feature was the geodesic frame construction of the fuselage by Barnes Wallis, the inventor of the skipping bomb of the 'Dam Busters' fame. The Wellington started the war as one of the principal bombers, only to be superseded by the heavies, but it was produced in greater numbers than any British-built bomber and for the duration of the war, quite a distinction. The Wellingtons operational history started on 4 September 1939 with the first bombing raid of the war on shipping at Brunsbüttel, the entrance to the Kiel Canal. A pair of Wellingtons were lost, becoming the first casualties on the Western Front. With the RAF change to night bombing, there were many Wellington losses. The Wellington participated in the first raid on Berlin on 25 August 1940. During the 1000-bomber raid on Cologne, Germany, on 30 May 1942, 599 of the aircraft were Wellingtons. The Wellington was also involved with Coastal Command in the Battle of the Atlantic. By 1944, it was the equivalent of an airborne command centre vectoring Mosquitos and Beaufighters on intercept missions.

Westland Welkin

The Welkin was a British twin-engine heavy fighter designed for attacks on possible extremely high-altitude Luftwaffe reconnaissance flights. The Welkin first flew on 1 November 1942. The power of the Merlins and the high-aspect wing enable the Welkin to climb to and fight at 43,000ft (13,106m) but it was restricted by compressibility problems due its long but thick wings. Few were built, and they only had a six-month operational life, which ended in November 1944.

The demand for the Merlin engine rose to a pitch as the late 1930s passed into the early 1940s. Manufacturers were clamouring for engines to suit the various types of aircraft being designed and their varied roles. Some aircraft required speed at low altitude, medium altitude, or high altitude, and some aircraft required longer range at those same altitudes. Some aircraft would be operating in extremes of climate such as Northern Russia and North Africa.

There was no possibility that Rolls-Royce Derby alone could cope with the demand, so two alternative manufacturing processes were initiated. They were the foreign manufacturers under licence and the 'Shadow' factories to increase the total production of the Merlin. Rolls-Royce, understandably, were reluctant, initially, to divulge the secrets of the Merlin but the exigencies of war overcame the caution. Rolls-Royce personnel were always present to give and gain knowledge as the manufacturing process suggested new engineering ideas.

SHADOW FACTORIES

As the possibility of another war became apparent, as early as 1934, the government set about a modest expansion of the RAF with the existing industry able to cope with the increase in required production. By 1935 the need for aircraft and aero engines had become more urgent and it was realised that the present manufacturers would not be able to meet the demand.

The proposed solution, named 'shadow scheme' internally in the government, called for shadow factories using existing companies involved in the motor car industry, with large-scale production experience, and building new factories. The 'shadow' refers to the fact that the work would be done by skilled motor industry people alongside, in the shadow of, their own motor industry operations and would allow for technology transfer to aircraft construction. The engines required by the RAF were made by Rolls-Royce, Bristol, Armstrong Siddeley and Napier.

The shadow scheme had two parts to it, the development of nine new factories, and expanding existing facilities to allow for an increase in production capacity or switching to the aircraft industry. The government would provide funding for the scheme, grants or loans, and perhaps the key to this whole scheme was Rolls-Royce and its plans for the Merlin engine. Bristol was also a main player with its Hercules engine but, interestingly, would not allow complete engines to be built offsite, just components. The first manufacturers proposed for the scheme were Austin, Daimler, Humber (Rootes Securities), Singer, Standard, Rover and Wolseley. The new factories were models of efficiency for their time. By 1938 the first bomber to be completely built at a shadow factory, Austin's, was flown a year ahead of the declaration of war. There would also be dispersal sites so as not to have 'all the eggs in one basket' in the event of aerial attack. At least Britain, eventually, had done some preparation ahead of the war, but would it be enough?

Rolls-Royce expanded the Derby factory between 1935 and 1939. In 1936 Ernest Hives, Managing Director of Rolls-Royce, was fully behind the shadow scheme, but he did have a few comments to make to the government. He wanted to emphasise that Derby was the centre of all development work with its skilled labour and that all work should remain there until all engineering bugs were removed from the first few hundred production engines. These Derby-trained skilled technicians would be moved to the shadow factories to ensure quality control of the Merlin product and provide manufacturing assistance. The Rolls-Royce shadow factory at Crewe, built 1938/39, on the other hand, was planned differently. Its workers were mainly unskilled and semi-skilled and so all planning, changes and modifications would be done at Derby and then the completed Merlin engine would be sent to Crewe for manufacturing. Also, due to lack of facilities, Crewe relied more on subcontractors for the engine parts. Hives advocated a factory at Glasgow.

In the end the Merlin was built at four locations in Britain and two, one postwar, in the USA. There were Rolls-Royce shadow factories at Crewe, Cheshire, and at Hillington near Glasgow, in Scotland. Ford built its shadow factory at Trafford Park, Manchester. In the US, Packard had its plant in Detroit, Michigan and, postwar, Continental Motors at Muskegon, Michigan. Ford initially had problems with the Rolls-Royce manufacturing tolerances – too much! Ford redrew the drawings at some cost in time but in the final analysis it was worth it for Ford achieved a high production rate, 200 per week. Previously Ford engineers had been at Derby for nine months learning the process. The Manchester factory was specifically tooled for the Merlin and produced a cheaper engine of no less quality. It would have difficulty switching to another engine type.

According to Peter Pugh in his book *The Magic of a Name: The Rolls-Royce Story*, there was a total of 168,040 Merlin engines built as follows: Derby 32,377, Crewe 26,065, Glasgow 23,647, Ford Manchester 30,428, Packard and Continental 55,523. To achieve this level of production Packard's employees went from 4,000 in the mid-1930s to 47,000 by 1945. These Merlins were installed in Hurricanes, Spitfires, Mosquitos, Lancasters and Mustangs and the list given earlier in this chapter.

A closer look at establishing the Hillington factory explains the challenges of establishing a new factory in a wartime environment. Peter Sherrard's book *Rolls Royce Hillington: Portrait of a Shadow Factory* published by the Rolls-Royce Heritage Trust is a good source of information on Hillington. Hives would have taken some of the following considerations into account when he was looking for a shadow factory location: One that was removed from the main areas of hostilities, a site that was large enough and could easily be developed, ideally close to rail and road transportation with options for air and sea transport, a skilled work force, if possible, with, most importantly, somewhere to house them (Penilee) close to the site, access to a source of raw material such as steel and the opportunity to subcontract parts in the local area. In January 1939 Rolls-Royce managers travelled north to interview companies that had replied to a Rolls-Royce newspaper advertisement looking for subcontractors. The replies and interviews were successful, and the go-ahead was given to start construction. By May, parts were being inspected at a temporary office on the site, before being sent to Derby, and by September 1939 the first two blocks of the factory were finished next to the Hillington 'Garden' Industrial Estate. A further twelve blocks were completed on the 100-acre site (40 hectares), which continued to have flooding problems due a small stream running through it.

Stow College in Glasgow established a Rolls-Royce training programme to train workers, male and female, on the many trades and on the use of machine tools. With further training at the Crewe Works, the pipeline of trained workers had begun deliver. By 1940 the foundry, die casting, pattern shop, and the rest of the factory were open for business. The first engine came off the line in October 1940 at the end of the Battle of Britain. Thousands of employees had worked two 12-hour shifts, seven days a week, to accomplish this feat. We need to also remember that the Hillington organisation was sending parts south to Crewe. In addition, the organisation had undertaken the repair and conversion of Merlin engines from the RAF.

The US continued to supply Britain with material assistance, although maintaining its neutrality until December 1941. In September 1941 there were 15,000 employees producing 98 per cent of the engine domestically. That was in contrast to Crewe and Derby, which obtained 50 per cent of their parts on a sub-contract basis. Of the employees 3,000 were female.

The 'Packard line' was established to strip, inspect, and, if necessary, upgrade, all engines coming from the US to meet the policy that only engines with the highest modification standard would be supplied to the RAF and the Allies.

Eventually the requirements outgrew the existing buildings and work began to be subcontracted. At the peak, 25,000 workers were under the direct control of Hillington, with 45,000 workers engaged in the Scottish aero industry. It is interesting to note that the repair, and the Packard upgradings and conversions, took up more time than new engines.

The work week gradually decreased from 82 hours to 47 hours a week after Hillington had reached the target of 100 engines a week in March 1942. A few bombs did drop on Hillington so workers taking to the air raid shelters in the middle of the night was not uncommon. The Hillington shadow factory was a huge success owing to the Rolls-Royce leadership and the dedication of the workers of the local area. The production figures speak for themselves but do not fully express the sacrifices and effort of the workers who made the Merlin the 'Engine that Won the Second World War'.

Examples of Factories

Location	Manager for Ministry of Aircraft Production	Original use	Wartime production
Acocks Green, Birmingham	Rover Aero	Westwood family's market garden	Parts for Bristol Hercules radial engine
Banner Lane, Coventry	Standard Aero No. 2	Golf course	Bristol Hercules sleeve valve radial engines
Blythe Bridge, Staffordshire	Rootes Securities		Blenheim, Beaufort, Beaufighter
Bolton, Lancashire	de Havilland		Propellers
Browns Lane, Coventry	Daimler	Farmland	Aero engines, Aircraft sub-assemblies
Burtonwood, Warrington	Fairey Aviation		Assemble and modify imported American aircraft
Canley-Coventry	Standard Aero No. 1	Vacant land Standard's Canley site	Bristol Beaufighter de Havilland Mosquito
Canley-Coventry	H. M. Hobson	Vacant land on Standard's Canley site	Carburettors for aircraft engines
Castle Bromwich, West Midlands	Nuffield Organisation then Vickers	Farm/Sewage works	Supermarine Spitfires, Avro Lancaster
Christchurch, Hampshire	Airspeed		Airspeed Oxford
Cofton Hackett, Longbridge	Austin	Farmland in Groveley Lane	Aero engines, Bristol Mercury and Pegasus Aircraft production – Fairey Battle, Stirling, Avro Lancaster, Wellington Bombers
Coventry, Stoke	Humber		Aero engines
Crewe, Cheshire	Rolls-Royce	Farmland	Rolls-Royce Merlin
Cwmbran, South Wales	Lucas	Farmland	Aircraft turrets

Location	Manager for Ministry of Aircraft Production	Original use	Wartime production
Errwood Park, Stockport	Fairey Aviation		Beaufighters then Halifax bombers
Drakelow Tunnels, Kidderminster	Rover Company	Hills	Machining parts for Bristol Hercules and Rolls-Royce Merlin
Hillington, Glasgow	Rolls-Royce	Farmland	Rolls-Royce Merlin
Meir, Stoke-on-Trent	Rootes Securities	Air Field	Harvard assembly, Mustang modifications
Ryton, Coventry	Humber	Farmland	Aircraft engines
Lode Lane, Solihull	Rover	Farmland	Parts for Bristol Hercules radial engine
Speke Airport, Lancashire	Rootes Securities	Speke Airport	Bristol Blenheim, Handley Page Halifax
Staverton, Gloucestershire	Rotol	Staverton Airport	Variable pitch propellers
Trafford Park, Manchester	Ford	Derelict motor assembly plant	Rolls-Royce Merlin
Oldham, Lancashire	H M Hobson	Cotton mill	Carburettors for aircraft engines

PACKARD

Regarding the decision to manufacture overseas, why the USA and why Packard? The manufacturing location had to be away from the current war zone as well as the foreseeable future war zone in case the battle was not going well. This would require the location far away from Germany and Italy in Europe. The location also had to have the resources, technical manpower, financial assets, political stability, manufacturing capability and the will to assist the Allies. Hence North America was the first choice, Australia being too far away and difficult in time and distance to get the engines and aircraft to Britain. The American Ford Motor Company of Dearborn, Michigan, in a suburb of Detroit, already had history with Rolls-Royce through the attempt to build Merlins at the Fordair plant, a subsidiary of the American Ford Motor company, in France, in 1940. The plan never materialised to any manufacturing extent before France was overrun by Germany in May 1940. Subsequently, there was a connection established through the Merlin Ford shadow factory in Trafford Park, Manchester, which by June 1941 was producing Merlin engines.

The Rolls-Royce representatives met with General Arnold and William Knudsen, Chairman of the General Motors Corporation, once the American government gave its approval to licensed production of the Merlin engine. Knudsen, as well as being close to President Roosevelt, headed the Office of Production Management. Connections are everything and there was no doubt that the Merlin would be built by someone in America. The first choice was the Ford Company, but it was not to be. Henry Ford was sympathetic towards Germany and was anti-Semitic in outlook; he was not alone in those views. The Ford Company continued to manufacture in Germany with their subsidiary, Ford-Werke, after the start of hostilities. Forde-Werke was associated with the practice of 'slave labour' before the separation from the parent company in America. Henry Ford refused to build the Merlin engine to support the war effort. The Japanese attack on Pearl Harbor, Hawaii, on 7 December 1941 changed all that. Ford threw his weight behind the American response and built a large aircraft manufacturing factory at Willow Run near Detroit, which produced the Consolidated B-24 Liberator. The largest assembly line in the world produced 9,000 B-24s, one coming off the line every 58 minutes, 18 hours from the start of its construction life; a demonstration of 'Fordism', mass production at its finest.

In June 1940, after the Ford turndown, Knudsen had talked with the Packard Motor Car Company and it agreed to build the Merlin. Packard was a good choice as it had previously served the country well during the First World War and had the manufacturing prowess required. Packard was founded by the two Packard brothers, James and William, and their partner George Weiss, who had invested in Winton cars. The Packard brothers believed they could improve on the Winton and by 1899 were building their own cars as the Ohio Automobile Company in Warren, Ohio. In 1902 the company was renamed the Packard Motor Car Company and moved to the East Grand Boulevard, in Detroit, in 1903. The factory and the Car Proving Grounds at Utica, Michigan, were designed by Albert Kahn, who was the foremost American industrial architect of his day. He is sometimes called the 'architect of Detroit'. The Packard automobile ended its production in 1956. When did it get involved with the aviation industry? With the Liberty L-12 engine.

Two more brothers would feature in Packard history, the Vincent brothers. Jesse studied engineering and worked for the Burroughs Corporation, where he excelled at finessing the production of the adding machines and registering many patents. He had always repaired tools on the home farm and he had an interest in automobiles. In 1912 he left Burroughs and went to Packard where, three years later, he designed his famous 'Twin-Six' automobile engine. He was a pilot, as well as driving Packard-powered speedboats. His brother, Charles, worked for the Hudson car company and invented the variable venturi carburettor. He too ended up at Packard. He was an avid competitive shooter and radio ham.

When the US declared war on Germany, the Aircraft Production Board summoned Jesse Vincent and Elbert Hall, of Hall-Scott Motor Co., to produce an engine equal to, or better than, the ones available to the Allies. The Liberty engine was tested on August 1917. The Liberty L-12 was an American 27-litre (1,649 cubic inch) water-cooled 45° V-12 aircraft engine of 400hp (300kW) designed for a high power-to-weight ratio and ease of mass production. More than 20,000 Liberty engines were built by the major car companies, Packard included. Jesse was commissioned as a Lieutenant Colonel in the Air Service to ensure standardisation. He established the Engineering Section at McCook Field in Dayton, Ohio. This was the beginning of Wright-Patterson Air Force Base and Materiel Command. After the war the Liberty engine was used as a speedboat engine. Packard went on to build a record-setting aviation diesel engine. Packard had proved it did have the attributes to be a player in the aviation engine business.

General Motors' Allison Division's V-1710 was the only available engine of note on short notice and its shortcomings were apparent with the Curtiss P-40. The Merlin engine version did improve the higher altitude performance of the P-40 but, as discussed in the previous chapter on the Mustang, it took the sleeker Mustang to finally make the engine/aircraft combination sing. In June 1940 an order was made to produce the Merlin XX as a 'stop-gap' measure. Packard was awarded a contract for 9,000 Merlin engines, 3,000 for the Air Corps and 6,000 for the British contract. In August a Rolls-Royce representative arrived to brief the Packard Company that there would be two Merlins manufactured by them and one of them would have a new two-piece block, still in the design phase. As an example, the Merlin XX became the Merlin 28 in Britain and the V-1650-1 in the USA. Information was missing from the British drawings as craft expertise and oral instructions were used in the factory. Because of the tolerances, each engine could probably be termed 'bespoke'.

The screw threads in the British engines were quite a collection of British Association, British Standard Fine, British Standard Whitworth and British Standard Pipe, both in left-handed and right-handed screws.

Imagine the planning and logistics of producing all these screws and bolts for the British engines. Packard used American Standard screws on the AAF Merlins. The first Curtiss P-40F flew on 30 June 1941, with some improved performance. In January 1942 everything had changed post-Pearl Harbor, in effect General Arnold told the British that they could have every engine over the '9,000' allocated to the Air Corps! What happened to the agreement of 6,000 engines? By the summer of 1942 Packard had developed its own two-speed, two-stage supercharger for the V-1650-3. This engine was the game changer and when installed in the Mustang P-51B the third contract changed for Packard, 13,325 V-1650-3s were ordered in June 1943. Some cooling problems with the now updraft carburettor (Allison was downdraft) were ironed out. The Merlin 65 was tried and fitted without too much modification to the airframe. The demand for the Merlin V-1650-3 far exceeded the capability of the Packard factory and that is a measure of its esteem in the high-performance world of the Second World War fighters.

These Packard engines did not differ sufficiently to merit a separate designation from the Rolls-Royce versions:

PACKARD	ROLLS-ROYCE	AIRCRAFT	INFORMATION
Merlin 28	Merlin XX R.M. 3 S.M.	Avro Lancaster III, Canadian Lancaster X	
Merlin 29	Merlin XX	Canadian built Hawker Hurricane	
Merlin 31	Merlin 21	Canadian built de Havilland Mosquito	
Merlin 33	Merlin 23	Canadian & Australian built de Havilland Mosquito	
Merlin 38	Merlin 22	Avro Lancaster III, Canadian Lancaster X	
Merlin T38	Merlin 22	Avro Lancaster III, Canadian Lancaster X	Modified Packard 38
Merlin 68	Merlin 85 R.M. 10 S.M.	Avro Lincoln II	Similar to V-1650-7
Merlin 69	Merlin 67	Canadian & Australian built de Havilland Mosquito	
Merlin 224	Merlin 24 R.M. 3 S.M.	Avro Lancaster III, Canadian Lancaster X	Single stage, two-speed

PACKARD	ROLLS-ROYCE	AIRCRAFT	INFORMATION
Merlin 225	Merlin 25	Canadian & Australian built de Havilland Mosquito	
Merlin 266-P	Merlin 66 R.M. 10 S.M.	Supermarine Spitfire XVILF	Two-stage, two-speed
Merlin 300	Merlin 100 R.M. 14 S.M.	Avro Lincoln	Speed density carburettor
Merlin 301	Merlin 100 R.M. 14 S.M.	Avro Lincoln	Speed density carburettor Reversed flow coolant

These Packard engines differed sufficiently to merit a separate designation from the Rolls-Royce versions:

PACKARD	ROLLS-ROYCE	TAKEOFF POWER	COMBAT POWER	AIRCRAFT	INFORMATION
V-1650-1	Merlin XX			P-40F Warhawk	Single stage, two-speed Merlin 28
V-1650-3	Merlin 61	1,380bhp @3,000rpm 1029kW	1600bhp @3,000rpm 1193kW 67" hg low blower	P-51B/C/F Canadian Mk 20	Critical altitude 11,800ft (3,597m) Merlin 63 two-stage
~~V-1650-5~~				P-63 Kingcobra	Not produced
V-1650-7	R.M. 10 S.M.	1,490bhp @3,000rpm 1111kW	1,590bhp @3,000rpm 1187kW 61" hg low blower	P-51D/K Canadian Mk 21	Critical altitude 8.500ft (2,591m) two-stage
V-1650-9* V-1650-9A	R.M. 16 S.M.	1,380bhp @3,000rpm 1029kW	1,930bhp @3,000rpm 1439kW 80" hg low blower (wet)* 10,100ft (3,079m)	-9 P-51H -9A P-51M	Beefed up engine to handle the wet power of the -9. -9A no water injection.

PACKARD	ROLLS-ROYCE	TAKEOFF POWER	COMBAT POWER	AIRCRAFT	INFORMATION
V-1650-11*	R.M. 16 S.M.	1,380bhp @3,000rpm 1029kW	2,150bhp @3,000rpm 1603kW 80" hg low blower (wet)* 7,700ft (2,347m)	P-51L cancelled due VJ day XP-82	Modified fuel system
V-1650-13					Not produced
V-1650-15					Not produced
V-1650-17					Not produced
V-1650-19					Not produced
V-1650-21				XP-82	Similar to V-1650-11 Left hand prop rotation
V-1650-23				P-82B/C/D	Similar to V-1650-11 Pressure carburettor
V-1650-25					Same as V-1650-23 Left hand prop rotation

*The wet power or water injection system used on the V-1650-9 and V-1650-11 engines is an anti-detonation injection (ADI) system, which allowed war emergency power settings to be used for a short duration. At high power settings the fuel-air mixture is greatly enriched to maintain coolant temperature within acceptable limits. The pressure injection carburettor would automatically adjust this rich mixture during high power requirements. This was quite easily discernible by a trail of black smoke emitting from the exhaust. To sustain engine efficiency without excessive fuel consumption or overheating, a water and methanol mixture was injected into the mixture at high power settings to substitute for the required extra fuel. The vaporised water assists to keep the combustion temperatures cooler, while the metering device maintains the best mixture for the demanded power. Methanol was added to the water to prevent the solution from freezing at cold temperatures. The P-51H ADI system contained enough solution for 7 minutes' operation.

It should also be mentioned that the Packard method of measuring engine boost was quite different from the Rolls-Royce method. All production Packard engine NAA Mustangs used a manifold pressure indicator calibrated for inches of mercury (Hg). The British system used pounds (lbs) of boost with the indicator showing zero for field barometric pressure. Boost was positive (+) when greater than barometric pressure and minus (-) when less than barometric pressure. The conversion factor is 1 pound of boost is equal to 2.04 inches of mercury. The combat power for the V-1650-3 is 67" Hg or +18lbs (67/2.04=32.8 minus atmospheric pressure 14.7 = +18.1) The major difference between the Rolls-Royce and Packard two-stage Merlins was the supercharger drive. Packard changed the Rolls-Royce Farman drive to their own designed epicyclic gearing.

What improvements changed the Merlin engine from a good engine to a great engine? There were many, many, improvements as the engine developed but two of note were by Schilling and the Ellor/Hooker engineering working partnership. Unlike fuel injection, which works under any 'G' (the force of gravity) conditions, positive or negative, the carburettor has problems with negative 'G' conditions. This is normally not a consideration during normal flight conditions but in the world of aerial combat, it is monumental. Instead of being able to bunt, put the nose down to rapidly dive the aircraft, it would have to be rolled to maintain positive 'G' and then, while upside down, the nose would be pulled to start descent. Valuable seconds lost in an evasive manoeuvre, with a fighter on your tail, could result in the aircraft being damaged or shot down.

This limitation was caused by the float type carburettor. A rapid dive would cause the float chamber to flood and the engine would splutter and misfire. A partial solution was found by Tilly Schilling, an engineer at the Royal Aircraft Establishment (RAE), Farnborough; using her ingenuity she developed a washer with a small hole in it to be installed in the fuel line that allowed just enough fuel for full power applications under negative 'G' conditions at sea level. Bendix in the USA cured the problem completely with the Bendix pressure diaphragm carburettor. Until Britain received the Bendix carburettor Rolls-Royce, RAE and SU developed anti-'G' versions of the Merlin carburettor.

The biggest and most noteworthy improvement was to the supercharger, which itself went through development in speeds and stages, that took the Merlin to new heights of power and which, in turn, brought the Merlin-powered aircraft to new levels of performance. The supercharger expert in Britain during the 1920s was James Ellor, who worked at the RAE.

Ellor was being head-hunted by the USA aviation industry to take his expertise there and RAE could not match the salary being offered as they were restricted by the Civil Service pay scale. To keep Ellor in Britain, the agreement was that Rolls-Royce would offer him employment. He joined Rolls-Royce in 1927 and among his early accomplishments was the supercharger on the 'R' engine of Schneider Trophy fame. Subsequently, the Merlin benefitted from this work. In 1940 he proceeded to Packard in the USA, and his contribution was recognised as outstanding. In 1944 he returned to Rolls-Royce and assumed his previous position as Chief Experimental Engineer.

During 1938 Stanley Hooker joined Rolls-Royce. He was hired primarily for his mathematical ability, he also had won the Busk Award in aeronautics in 1928 at the Imperial College, London. He was virtually ignored during his first few days at work, on purpose as it later turned out, and while wandering around the factory he discovered the supercharger section where he thought his expertise and knowledge could be used.

The power of an engine depends on the mass of air and fuel that it can consume, and the supercharger provides the means of getting the maximum amount into the engine. It does this by a fast, for example 28,000rpm, rotating rotor with radial vanes to distribute the air centrifugally, which is then collected in a diffuser that routes the pressurised air to the cylinders. The total efficiency depends on the efficiency of the parts, the rotor and diffuser, and Hooker discovered that the Merlin was only 65 per cent efficient and could be improved to 75 per cent efficiency using his new calculations. On seeing the report, Ellor immediately put him in charge of supercharger development.

The supercharger evolved through the single speed, single stage to the two-speed, single stage and finally to the two-speed, two-stage with its intercooler to keep the air mixture cool between stages. It was this last Hooker development that turned the corner in the performance of the British fighters and bombers and turned the Mustang into the great fighter escort with its Merlin 68 or Packard V-1650-3 two-speed, two-stage Merlin engine. Notwithstanding all this information on the Merlin engine and its very important supercharger, the Merlin did have other applications, in tanks with an un-supercharged engine based on the Merlin called the Meteor engine. This engine had to run on the 'pool fuel' of battlefield conditions. In April 1941 the first Crusader tank with a 600bhp (450kW) engine was field-tested and subsequent production models were called the Cromwell and built by the Birmingham Railway Carriage and Wagon Company.

The Merlin

The Merlin was built in approximately seventy different variants by Rolls-Royce and Packard. Each of the more than 168,000 engines were paired to a varied requirement from fighter to bomber to fighter bomber to fighter escort at low, medium and high altitudes in all parts of the world, both hot and cold. What a legend and what a legacy; so it is no wonder that the Merlin is lovingly preserved and restored approaching ninety years since it was first a concept in the mind of Henry Royce. The sound of a Merlin in full flight can be experienced at many airshows; ah, that sound, once heard, never forgotten.

Model	Take-off Power @3,000 RPM	Combat Power	Aircraft	Information
PV-12	740hp 552kW		Hawker Hart	Initial design used an evaporative cooling system. Two built, passed bench Type Testing in July 1934. First flown 21 February 1935.
Merlin B	950hp 708kW			Two built, ethylene glycol liquid cooling system. "Ramp" cylinder heads (inlet valves were at a 45-degree angle to the cylinder). Passed Type Testing February 1935.
Merlin C	950hp 708kW		First Flight Hawker Horsley 21 December 1935	Development of Merlin B. Three separate castings, crankcase and cylinder blocks. Bolt-on cylinder heads.
Merlin E	950hp 712kW	1,045hp 779kW	Supermarine Spitfire prototype	Similar to C, minor design changes. Passed 50-hour civil test in December 1935 but failed military 100-hour test in March 1936.
Merlin F			Hawker Horsley, test flight	Similar to C, E. First production ramp head engine; unsuccessful, few made. Designated Merlin I
Merlin G		1,030hp 768kW		Ramp heads replaced with parallel valve pattern. Designated Merlin II
Merlin I	890hp 664kW		Battle Mk.I	First production Merlin; 172 built. Merlin I to III used 100% glycol coolant.
Merlin II (RM 1S)	880hp 5,500ft 656kW 1,676m	1,030hp 768kW	Spitfire Mk.I, Defiant Mk.I, Hurricane Mk.I, Sea Hurricane Mk.I, Battle Mk.I	Used 100% glycol coolant. First production Merlin II delivered 10 August 1937.

Model	Take-off Power @3,000 RPM	Combat Power	Aircraft	Information
Merlin III (RM 1S)	880hp 656kW	1,310hp 9,000ft 977kW 2,743 m	Spitfire Mk.I, Defiant Mk.I, Hurricane Mk.I, Sea Hurricane Mk.I, Battle Mk.I	Variant of Merlin II with universal propeller shaft, de Havilland or Rotol propellers. Late 1939, used 100 octane fuel. First production Merlin III delivered 1 July 1938.
Merlin VIII	1,080hp 805kW	1,275hp 951kW Sea Level	Fulmar Mk.I	100 octane fuel.
Merlin X (RM 1SM)	1,280hp 954kW	1,280hp 954kW Sea Level	Halifax Mk.I, Wellington Mk.II, Whitley Mk.V and Whitley Mk.VII	First production Merlin to use a two-speed supercharger, 5 December 1938.
Merlin XII (RM 3S)	1,170hp 876kW	1,280hp 10,500ft 954kW 3.200m	Spitfire Mk.II	Coffman cartridge starter. First Merlin to use 30/70% glycol/water coolant. First production Merlin XII, 2 September 1939.
Merlin XX (RM 3SM)	1,280hp 954kW	1,490hp 12,500ft 1,111kW 3,810m	Beaufighter Mk.II, Defiant Mk II Halifax Mk.II, Halifax Mk.V, Hurricane Mk.II and Hurricane Mk.IV, Lancaster Mk.I, Lancaster Mk.III, Spitfire Mk.III	First production version with two speed supercharger, 4 July 1940.
Merlin 21	1,280hp 954kW	1,490hp 12.500ft 1,111kW 3,810m	De Havilland Mosquito Mk.I, Mk.II, Mk.III, Mk.IV and Mk.VI	Merlin XX with direction of coolant flow reversed for Mosquito wing radiator installation
Merlin 22	1,390hp 1,037kW	1,435hp 11,000ft 1,070kW 3,353m	Lancaster Mk.I, York Mk.I	Merlin 22A is Merlin XX converted to two-piece cylinder blocks Merlin 21 converted to two-piece cylinder blocks

Model	Take-off Power @3,000 RPM	Combat Power	Aircraft	Information
Merlin 23	1,390hp 1,037kW	1,435hp 11,000ft 1,070kW 3,353m	De Havilland Mosquito Mk.I, Mk.II, Mk.IV, Mk.VI, Mk.XII and Mk.XIII	Merlin 22 with direction of coolant flow reversed for Mosquito wing radiator installation Merlin 23A Merlin 21 converted to two-piece cylinder blocks
Merlin 24	1,610hp 1,201kW	1,510hp 9,250ft 1,126kW 2,819m	Lancaster Mk.I, Lancaster Mk.VII, York Mk.I and Halifax Mk.II [14]	Merlin T-24-1,2,3,4 Transport Command
Merlin 25	1,610hp 1,201kW	1,510hp 9,250ft 1,126kW 2,819m	De Havilland Mosquito Mk.VI and Mk.XIX	Merlin 24 with direction of coolant flow reversed for Mosquito wing radiator installation
Merlin 27	1,610hp 1,201kW	1,510hp 9,250ft 1,126kW 2,819m	Hurricane Mk.IV	
Merlin 28	1,300hp 969kW	1,240hp 11,500ft 925kW 3,505m	Lancaster Mk.III, Kittyhawk II (Curtiss P-40F)	Built by Packard as the **V-1650-1**
Merlin 29	1,300hp 969kW	1,240hp 1,500ft 925kW 3,505m	Hurricane Mk.XII (Canadian-built), Kittyhawk II (Curtiss P-40F)	Splined propeller shaft
Merlin 30	1,300hp 969kW	1,360hp 6,000ft 1,014kW 1,829 m	Barracuda Mk.I and Fulmar Mk.II	
Merlin 31	1,300hp 969kW	1,240hp 11,500ft 925kW 3,505 m	Mosquito Mk.XX (Can), Mosquito Mk.40 (Aus), Kittyhawk II (P-40F and L)	Packard **V-1650-1**
Merlin 32 (RM 5M)	1,620hp 1,208kW	1,640hp 2,000ft 1,223kW 610m	Barracuda Mk.II, Seafire Mk.II, Hurricane Mk.V, Spitfire PR Mk. XIII	A "low altitude" version of Merlin with cropped supercharger impellers for increased power at lower altitudes, used mainly in Fleet Air Arm aircraft, 17 June 1942.

Model	Take-off Power @3,000 RPM	Combat Power	Aircraft	Information
Merlin 33	1,400hp 1,044kW	1,400hp 11,500ft 1,044kW 3,505 m	Mosquito XX (Canadian), Mosquito 40 (Australia)	**Packard-built Merlin 23**
Merlin 38	1,400hp 1,044kW	1,400hp 11,500ft 1,044kW 3,505 m)	Lancaster. I and II	**Packard-built Merlin 22**
Merlin 45 (RM 5S)	1,185hp 884kW	1,515hp 11,500ft 1,130kW 3,505 m	Spitfire Mk.V, Spitfire PR Mk.IG (later redesignated Spitfire PR.VII), Spitfire PR Mk.IV, Seafire Mk.IB, Seafire Mk.IIC	Single-stage, single-speed supercharger for low altitude Spitfire use, 13 January 1941. First of specialised engines for Spitfire Mk V variants and early Seafires. Based on XX engine.
Merlin 45M	1,230hp 917kW	1,585hp 2,750ft 1,182kW 838 m	Spitfire LF Mk.V	Based on Merlin 45 with "cropped" supercharger impeller allowing greater boost at low altitudes.
Merlin 46	1,100hp 820kW	1,415hp 14,000ft 1,055kW 4,267 m	Spitfire Mk.V, Spitfire PR Mk.IV, Spitfire Mk.VII, Seafire Mk.IB and Seafire Mk.IIC	Merlin 56 is same except diaphragm-controlled fuel feed in carburettor
Merlin 47 (RM6S)	1,100hp 820kW	1,415hp 1,400ft 1,055kW 427 m	Spitfire HF Mk.VI high-altitude interceptor	Marshall compressor (blower) to pressurise the cockpit, 2 December 1941.
Merlin 50 (RM5S)	1,185hp 884kW	1,470hp 9,250ft 1,096kW 2,819 m	Spitfire Mk.V	Low-altitude version with supercharger impeller "cropped". Bendix Stromberg "negative-g" carburettor. Merlin 50 with Merlin 46 rotor and guide vanes
Merlin50M (RM 5S)	1,230hp 917kW	1,585hp 2,750ft 1,182kW 838 m	Spitfire LF Mk.V	
Merlin 55	1,185hp 884kW	1,470hp 9,250ft 1,096kW 2,819 m	Spitfire Mk.V and Seafire Mk.III	Merlin 55A similar to Merlin 45, lower modification standard

Model	Take-off Power @3,000 RPM	Combat Power	Aircraft	Information
Merlin 55M	1,230hp 917kW	1,585hp 2,750ft 1,182kW 838 m	Spitfire LF Mk.V, Seafire Mk.III	Variant with "cropped" supercharger impellor Merlin 55MA is Merlin 45 with two-piece cylinder block
Merlin 60	1,390hp 1,037kW	1,110hp 29,000ft 828kW 8,839 m	Wellington Mk.VI	First variant fitted with two-stage, two-speed supercharger; rated for high altitude.
Merlin 61 (RM8SM)	1,280hp 954kW	1,565hp 11,250ft 1,167kW 3,429 m	Spitfire Mk.IX and Spitfire PR Mk.XI	Two-speed two-stage supercharger providing increased power at medium to high altitudes. Incorporated two-piece cylinder blocks designed by Rolls-Royce for the Packard Merlin 2 March 1942.
Merlin 62	1,390hp 1,037kW	1,110hp 29,000ft 828kW 8,839 m	Wellington Mk.VI	
Merlin 63	1,280hp 954kW	1,710hp 8,500ft 1,275kW 2,591 m	Spitfire Mk.VIII, IX, PR.XI	Replaced Merlin 61
Merlin 63A	1,280hp 954kW	1,710hp 8,500ft 1,275kW 2,591 m	Spitfire PR Mk.XI	
Merlin 64	1,280hp 954kW	1,710hp 8,500ft 1,275kW 2,591 m	Spitfire Mk.VII	Cabin pressure blower
Merlin 65	1,315hp 981kW	1,705hp 5,750ft 1,271kW 1,753 m	Mustang- project only	Similar to Merlin 63 with Bendix-Stromberg carburettor
Merlin 66 (RM 10SM)	1,315hp 981kW	1,705hp 5,750ft 1,271kW 1,753 m	Spitfire LF Mk.VIII LF Mk.IX	Fitted with supercharger rated for low altitude; Bendix-Stromberg anti-g carburettor

Model	Take-off Power @3,000 RPM	Combat Power	Aircraft	Information
Merlin 67	1,315hp 981kW	1,705hp 5,750ft 1,271kW 1,753 m	Mosquito	Similar Merlin 66, reverse flow
Merlin 68	1,670hp 1,245kW	1,710hp 6,400ft 1,275kW 1,951 m	Mustang III (North American P-51B and C) Lincoln	**Packard version of Merlin 85**
Merlin 69	1,670hp 1,245kW	1,710hp 6,400ft 1,275kW 1,951 m	Mustang III and Mustang IV (North American P-51C, D, F and K) Mosquito	**Packard version of Merlin 67**
Merlin 70	1,250hp 932kW	1,655hp 10,000ft 1,234kW 3,048 m	Spitfire HF Mk.VIII HF.IX, PR.XI	Bendix Stromberg anti-g carburettor
Merlin 71	1,250hp 932kW	1,655hp 10,000ft 1,234kW 3,048 m	Spitfire HF Mk.VII	Cabin pressure blower, Bendix Stromberg anti-g carburettor
Merlin 72	1,280hp 954kW	1,710hp 8,500ft 1,275kW (2,591 m	De Havilland Mosquito PR Mk.IX, B Mk.IX, Mk.XVI and Mk.30. Welkin Mk.I	Merlin 63 reverse flow coolant
Merlin 73	1,280hp 954kW	1,710hp 8,500ft 1,275kW 2,591 m	De Havilland Mosquito Mk. XVI, Welkin Mk.I	Same as Merlin 72 with a cabin pressure blower
Merlin 76 (RM 16SM)	1,250hp 932kW	1,655hp 10,000ft 1,234kW 3,048 m	De Havilland Mosquito PR Mk.XVI, Mk.30, Welkin Mk.I	1,233 hp (919kW) at 35,000 ft (10,668 m); Dedicated "high altitude" version used in the Westland Welkin and some later Spitfire and de Havilland Mosquito variants. Bendix-Stromberg carburettor
Merlin 77 (RM 16SM)	1,250hp 932kW	1,655hp 10,000ft 1,234kW 3,048 m	De Havilland Mosquito Mk. XVI, Welkin Mk.I, Spitfire PR Mk.X	Same as Merlin 76 with a pressurising blower

Model	Take-off Power @3,000 RPM	Combat Power	Aircraft	Information
Merlin 85	1,635hp 1,219kW	1,705hp 5,750ft 1,271kW 1,753 m	Lancaster Mk.VI and Lincoln Mk.I	Merlin 85A/B,86 similar to Merlin 85 with improvements
Merlin 90	1,225hp 914kW		Tudor prototype	Similar Merlin 100 with two-speed, single stage supercharger
Merlin 102	1670hp 1245kW	1,810hp 1350kW	Tudor	Similar Merlin 100, Merlin 102A has after heater
Merlin104	1245kW1,670hp	2,030hp 1514kW	Mosquito	Similar to Merlin 114, lower gearing on supercharger
Merlin 113	1,535hp 1145kW	1,690hp 1260kW	Mosquito	Similar to Merlin 100, reverse coolant flow and high supercharger gearing. Merlin113A had anti-surge supercharger diffuser
Merlin 114	1,535hp 1145kW	1,690hp 1260kW	Mosquito	Similar to Merlin 113 with cabin supercharger drive. Merlin 114A with anti-surge supercharger diffuser.
Merlin 130/131	2,070hp 1,544kW		De Havilland Hornet F Mk.1, PR Mk.2, F Mk.3, FR Mk.4.	Slimline engine modified to decrease frontal area to a minimum and was the first Merlin series to use down-draught induction systems. 130 port/131 starboard engine
Merlin 132	1,670hp	2,030hp	Sea Hornet	Port engine
Merlin 133/134	1,670hp	2,030hp 1,514kW	De Havilland Sea Hornet F Mk.20, NF Mk.21 and PR Mk.22	Derated 130/131 Merlin 133 is port engine Merlin 135 is starboard engine

Model	Take-off Power @3,000 RPM	Combat Power	Aircraft	Information
Merlin 224	1,610hp 1,201kW	1,635hp 1219kW	Avro Lancaster Mk.III, Mk VI and Mk.X	**Packard-built Merlin 24** (2 prefixes are Packard built)
Merlin 225	1,315hp 981kW	1,635hp 1219kW	De Havilland Mosquito Mk.25 and Mk.26	**Packard-built Merlin 25**
Merlin 266 (RM 10SM)	1,315hp 981kW	1,705hp 1271kW	Spitfire LF Mk.XVI	**Packard-built Merlin 66**, optimised for low-altitude operation.
Merlin 620		1,795hp 1,339kW	Avro Tudor, Avro York, and Canadair North Star	The Merlin 620–621 series was designed to operate in the severe climatic conditions encountered in Canada and for long-range North Atlantic air routes.

PART 5

THE MERLIN LIVES ON
2018
Eighty-Five Years Young

10

Preserved and Restored

At the present time the Merlin can be found in many different conditions, from the very sad remnant to the majestic example of a fine piece of preserved engineering. Some engines remain as hidden treasures buried underground at home in Britain, in mainland Europe or elsewhere in the world. Some of these engines' locations are known to aviation enthusiasts and are awaiting excavation but some are not and never will be, lost to aviation history forever as the result of enemy action, flying accident or mechanical failure.

Some recognisable Merlin engines are being held by collectors for future purposes, just as they were discovered; some are being slowly made presentable with genuine parts by aviation enthusiasts, such as the one held by Martin Rouse, discussed later. Some have been restored for display and look as if they have just come out of the factory, but do not run. Some have been restored to running condition and are mounted on an engine stand to run at airshows and aviation displays. Some are still being restored.

Some engines have been restored to an operating condition and installed in aircraft that they were designed for, this is the realm mainly of aviation museums, and are used for ground running exhibits only. These engines can be seen and heard in Avro Lancasters at the Bomber Command Museum of Canada, Nanton, Alberta, Canada (see page 203) and the Lincolnshire Aviation Heritage Centre at East Kirkby, Lincolnshire, England.

Some engines have reached the pinnacle of homage to the great Merlin, they and the aeroplane have both been restored to a government-certifiable flying condition. Perhaps the most visible and famous examples of this are the Merlin-powered aircraft – Spitfire, Mosquito and Lancaster – of the Battle of Britain Memorial Flight, based at RAF Coningsby, Lincolnshire England. The Canadian Warplane Heritage Museum, Hamilton, Ontario, flies the Lancaster and other Merlin-powered types. The Flight and Museum are described over the following pages.

That is not to discount or forget the howl of the Merlins at the Reno Air Races, Nevada, USA. In addition to these engines, many private aircraft Merlins fly worldwide in various aircraft including a dual-seat Spitfire in England where you can enjoy a flight in the historic fighter. Hurricanes, Spitfires and Mustangs are common on the major airshow circuit but there are only two Lancasters flying in the world and, difficult to verify, but at last count there were less than a handful of Mosquitos flying. The sound of the 'Flying Legends' at the recent Imperial War Museum, July 2017 airshow at Duxford, Cambridgeshire, England, attests to the fact that the Merlin is still as popular as ever.

Notwithstanding the sights and sounds of the single Merlin engine in the Spitfire and Hurricane, and the rare occasion when you will experience two Merlin engines in the Mosquito, the world is very fortunate that there are, at the time of writing in the summer of 2018, several opportunities to hear four Merlins together in a sea of sound in the Lancaster. The sound of four Merlins flying overhead is thrilling enough but when you have the opportunity to stand in close proximity to them, that is spine-tingling for an aviation or Second World War aficionado.

There are four places that you can see a real live Lancaster during the extended summer months of the year. It is not surprising that they are in England and Canada, the two places that manufactured the Lancaster. What is surprising is that three of the places are privately run organisations trying to keep the memory of the Lancaster and Bomber Command alive. This is not to say that at times during the past, various levels of government have not contributed to the effort to keep these old warbirds running. Once witnessed it is hard to forget the sights and sounds of four Merlins in full song. There were twelve in total in 2014 as the Battle of Britain Memorial Flight and Canadian Warplane Heritage Museum Lancasters flew over the Lincolnshire Aviation Heritage Centre Lancaster on the field at East Kirkby, Lincolnshire.

The Lancaster is a part of the history and heritage of Canada and the British Isles and of the freedom we have been fortunate to enjoy these past seventy years. These four aircraft are a fitting tribute to Bomber Command and the crews that served in them. The two of them in Canada are named after Victoria Cross winners, S/L Ian Willoughby

Bazalgette, VC, DFC, for the Bomber Command Museum of Canada Lancaster, and Pilot Officer Andrew C. Mynarski, VC, for the Canadian Warplane Heritage Museum Lancaster. The Lincoln Aviation Heritage Centre Lancaster chose a famous wartime national newspaper cartoon character, 'Just Jane', who was a daily pin-up girl for the *Daily Mail*. (Fred Panton, one of the museum founders, had attended a show by Chrystabel Leighton-Porter impersonating Just Jane in a Skegness theatre.) The BBMF Lancaster has changed its colour scheme over the years to honour particular people or aircraft.

My own association with the Merlin came first during my air spotting days during the 1960s when I saw and heard the Dan Air of London Avro York and British Midland Canadair C-4 Argonauts. The next time would be at the 1968 Abbotsford International Airshow in Abbotsford, British Columbia, Canada, as a private Spitfire taxied in front of an RAF Vulcan prior to an air display. I have the 'slide' in my collection, quite a contrast in size and style. The next time I heard Merlins was when the four belonging to the Canadian Warplane Heritage Museum's Lancaster were part of the 2010 Abbotsford Airshow.

The following excerpts on the four museums are in part taken from my book *The Lancaster* published by Amberley Books in 2015 to give the reader some idea of the Merlin engine at work today. The book takes a reader on a comprehensive journey of the Lancaster, for those interested in further reading.

Battle of Britain Memorial Flight
www.raf.mod.uk/bbmf
RAF BBMF Visitors Centre, Dogdyke Road, Coningsby,
Lincolnshire, England LN4 4SY

Due to the fact that the Battle of Britain Memorial Flight Lancaster is administratively part of the RAF No. 1 Group, it is the most restrictive among the four museums for close access to the aircraft. Sadly, interviews and photographic opportunities are limited, due to the cost of insurance for liability concerns on an active RAF Station. PA474 can be seen as part of a guided hangar tour arranged by the Visitor Centre in co-operation with the Lincolnshire County Council. The public tours started in 1986 and more than 300,000 visitors have seen the Memorial Flight aircraft. The historic collection includes examples of the Supermarine Spitfire, Hawker Hurricane, Douglas Dakota and de Havilland Chipmunk.

However, the good news is that the Lancaster is highly visible during the display season. It is regularly seen at air displays, events commemorating the Second World War, and British State occasions, such as the Trooping the Colour celebrating Queen Elizabeth II's birthday. In 2014, for instance, it made 158 appearances not including the eighteen displays and twenty-nine flypasts with the Canadian Warplane Heritage Museum Lancaster during August and September. It takes a lot of effort to keep these seventy-year-old engines serviceable and every now and again there are problems, as illustrated by a BBMF news release dated 29 August 2017.

Following the recent pause in flying our Merlin-powered aircraft for safety reasons, OC (Officer Commanding) BBMF said: 'Please see below the MOD's (Ministry of Defence) latest release on the subject of BBMF Merlin engines. As I hope you all know, my aim is to keep these vital and unique artefacts of Britain's proud history in the sky forever. To do that we sometimes have to take extra precautions, with our eye sharply on continuing safety, to ensure that we can keep them flying for future generations. I thank each and every one of you who has voiced their support and shown their understanding. We will get them back into the blue as soon we can.' The update from the MOD: 'Our investigation has confirmed an issue with a pinion gear in a Merlin engine. With the precise cause of the problem known, each pinion gear is now being inspected to confirm it meets our exacting standards, with the BBMF and industry putting all of our efforts into getting these beautiful aircraft safely back in the air as soon as possible.'

Bomber Command Museum of Flight
www.bombercommandmuseum.ca
1729 21st Avenue, Nanton, Alberta, Canada T0L 1R0

My next association, indirectly, with the Merlin was when I wrote the *Lancaster Manual 1943* and *The Lancaster* books for Amberley Books and used the facilities of the Bomber Command Museum of Canada (BCMC) in Nanton, Alberta, Canada, for my research. The centrepiece of the museum is the Avro Lancaster that is restored for ground running only and features highly in their theme. The engine run-ups, both day and night, are held at various published times during the museum opening season. The invitation was straightforward enough. Dave Birrell, a Director of the Bomber Command Museum of Canada, who also wears a second hat as the Librarian and Archivist for the museum, invited me to be onboard the museum's Avro Lancaster for the first four-engine run-up of the 2014 season. The run-up on 24 April 2014 was to prepare the aircraft for the RCAF 90th Anniversary celebrations to be held that weekend.

The Friday in April was cool and dry as the aircraft was towed out of the hangar. Good weather for the Canadian prairie in spring. I stood and wondered what it would all mean to me. I was no stranger to combat machines, but it had been forty years since I had climbed into my own Avro, one of life's coincidences, CF100 fighter and flew secret electronic warfare missions for the North American Air Defence Command. I looked across at the black Lancaster as it sat there silent and formidable – just me and the 'Lanc' for a few moments of contemplative respect.

The Lanc with a crew of seven must have seemed so small and almost claustrophobic to the men as they climbed aboard with their parachutes and equipment and struggled in to their small work areas. On the other hand, it must have seemed so big a target to the Luftwaffe night fighters as they attacked from below and could see the aircraft silhouetted against the stars or searchlight beams of light.

An aircraft of contrasts – the bomb bay doors were open, and I could see the large bay which could carry bombs weighing up to 22,000lb (9,979kg) in size. I looked into the rear gunner's turret, small, and wondered how anyone could get in there, let alone operate the four guns. The aircraft had what I can only describe as an arrogant and aggressive stance as it sat there on the apron with its nose in the air. The four Merlins hung on the wings, cowlings painted a nice black, in silent splendour with their propellers all neatly aligned for the spectators. I had written a book on this aircraft and yet I understood, at that moment, that I did not know it at all until I experienced the sights and sounds of its four engines running; that would soon change.

The daylight was fading as I returned to the warmth and light of the hangar. I now waited, with some trepidation, to return to the Lancaster in the dark and share the excitement of the engine start. Would I fail as an author to capture the atmosphere onboard and identify with the young men who had risked their lives at appalling odds to carry out the raids of Bomber Command? I, indeed, have a heavy burden on my shoulders to tell their story and do justice to this magnificent aircraft, the Avro Lancaster.

Thanks to Dave and the BCMC I will have this experience of a four-engine night start to help me in my quest. The night was planned to be coincidental with a 'Thank the Volunteers' evening with food and libation on hand. Somehow, I thought of myself as an intruder and could not really partake in all the activities provided for the hard-working volunteers. I observed the men and women who had all worked hard the previous year to keep the museum not only viable but to plan future growth. Perhaps in some way I was a kind of volunteer as I would be writing about their Lancaster and promoting it and the BCMC. I enjoyed a plate of vegetables as I wandered around looking at the various exhibits, which included a working tail turret and a cutaway Merlin engine. I marvelled at the number of moving parts of the twelve-cylinder engine and appreciated the symmetry of the two banks of cylinders.

The time came to board the aircraft as there was a general movement in the direction of the apron outside the hangar doors. Slightly overdressed in a fleece and parka, after all it was April in Alberta and snow was in the forecast, I approached the front of the dark aircraft as it sat poised to come to life. Some introductions by Dave Birrell of the Aircraft Captain and Engineer, a quick briefing on exiting the aircraft if a fire occurs, and I was ushered to a set of steps underneath the escape hatch in the bomb aimer's compartment. The wartime crews would normally enter the aircraft door at the starboard rear of the aircraft. The Engineer for this night run-up was Greg Morrison.

Once again, I felt like an interloper in this alien world of the Second World War bomber crew. The war occurred before I was born. I certainly had never been near such an aircraft before, let alone being moments away from entering it. Clambering up the steep steps I then pulled myself on to the floor of the bomb aimer's compartment. I was guided to the steps at the rear of the compartment by a flashlight held by someone above who I could hear but only see his feet. A few lights illuminated my progress as I climbed out of the compartment and passed by the Pilot perched high above and the Flight Engineer's instruments on the starboard fuselage wall.

Squeezing past the navigator's table I was assigned to the radio officer's position just in front of the spar that blocked normal progress through the fuselage. It required a certain agility to get over the space between the spar and the fuselage roof – a vertical hand-hold bar assisted the manoeuvre. It certainly would be a tight fit, especially for seven men in winter flying gear and carrying their parachutes. A Bomber Command Museum of Canada volunteer staff photographer, Doug Bowman, was my companion for the engine start and worked very hard in the confined space to get images of the cockpit, engine instruments and engine exhaust.

In the darkness I heard voices responding to the checklist, which indicated the initiation of the start procedure. The checklists are completed by a challenge and response system and are heard by the spectators on loudspeakers, which really add to the atmosphere of the engine start.

PRESTART

1.	PARK BRAKE	SET
2.	BOOST COIL SWITCH	OFF
3.	MAGNETO SWITCHES	OFF
4.	MASTER FUEL COCKS	OFF
5.	THROTTLES	CLOSED

Prior to the actual start I heard some banging that echoed through the fuselage, perhaps something to do with the hydraulics or the fuel system? A few reselections of some circuit breakers, 'try the ones along the bottom,' and the first engine burst into life. That is all twelve cylinders of the Rolls-Royce Merlin engine. On this particular Lancaster there is a small window on the starboard side through which I could see the blue exhaust flames of number three engine. They danced haphazardly and hypnotically like the Northern Lights. I actually enjoyed the sound of the initial rough idle with its pops and backfires of exhaust flames before it settled down to a steady beat.

The second engine burst into life and Bowman got some good images of the two sets of engine exhausts illuminating the darkness. The noise by now had reached quite a din with two engines at idle. It was interesting that the cockpit voices, Pilot and Engineer, were played over the loudspeakers to the crowd outside so that they could follow the engine start. When the engine started next to my sitting position on the port side I thought that the propeller must surely come through the side of the fuselage. The tip of the propeller was separated from the crew, me on this night, by the thickness of a sheet of aluminium. I cannot imagine what the noise inside must have been like for a six-hour mission in spite of leather helmets with earphones.

The engines were revved up to 1,600rpm during the engine run-up as per the checklist written specifically for the BCMC's Lancaster by engineer Greg Morrison. The rpm has been increased to 1,800rpm for 2018 with the corresponding increase in noise for the spectators. The sound is like being in a tin can, which I literally was, caused by the deep exhaust roar of the four twelve-cylinder Merlins and accompanied by the shaking of the fuselage, quite the new life experience.

During the night run-up the propellers and aircraft were floodlit to great effect for the spectators; I missed that being inside. All too soon the engines were shut down to be replaced by the sound of the spectators cheering and clapping, another successful four-engine start due to the many volunteers who made it happen. Everything had gone according to plan, which bode well for the engine starts scheduled for the next day.

I carefully exited the aircraft and marvelled at how a crew of seven could work in such close quarters over long periods of time exposed to constant danger – a young man's game, for sure. Most of the crew would have been in their twenties which would be an advantage for the physical stress, but the mental stress would have been different as they had their full lives ahead of them, if they were lucky. I was humbled by my night-time run-up experience and I would like to thank Dave Birrell and the volunteers for this once in a lifetime experience. For me a 'four Merlin moment' as they continue their illustrious life, this one in peacetime.

Three years later I spoke to Greg Morrison at the end of their engine run-up season in 2017. It was later that year as they had a special day for the Canadian Aeronautical Preservation Association. The museum completed seventeen official run-ups during the season. A timing issue was the first challenge of the season following some engine work. The 2017 winter preparation starts with the engine cowls coming off the four engines and a thorough inspection of the engines for any obvious breaks in lines, fluid leaks, chafing wires and corrosion. The previous winter inspection had indicated a problem, the running symptoms confirmed by Vintage V12's of California, with the number one engine supercharger. As well as the museums and their staff there are other separate organisations that specialise in looking after the operational Merlin engines. The museums have limited capabilities and often subcontract the specialised repairs, overhauls and restoration to companies throughout the world. One of these companies is the well-known Vintage V12's.

The supercharger on the Packard-built Merlin 224, weighed 200lb (91kg) and required two men to remove and carry it, was sent to Vintage V-12's for rebuilding. This repair was an administrative financial decision as it cost thousands of dollars but was required to keep the number one engine serviceable for the 2017 season and maintain the four engine run-up scenario. The superchargers are locked in the low blower (MS) position for the run-up. There were also some timing issues, which the team solved in-house. The team, down to three volunteers, consists of lead Greg Morrison, Aircraft Maintenance Engineer, and two experienced helpers, Brian and Shane. Routine maintenance such as pulling the spark plugs and draining oil screens is accomplished.

The number three engine had some coolant leaks, which required some welding of the corrosion holes in the supply pipes. It started with the number four engine rubber mounting pads, the seventy-year-old pads had turned 'as hard as glass' Greg recounted. A local manufacturer was found to make suitable pads and all the engine pads were replaced with a new age material, Viblon. Canadian sources had overhauled the propellers in the past, Canadian Propeller in Winnipeg, Manitoba, and Aero Propeller in Calgary, Alberta. Not only does the museum preserve wartime history but in doing so it contributes to the local economy as much as possible. To save money the ignition harnesses, built in-house, and plugs are presently being slowly converted to automotive ones which are readily available, much cheaper, and performing well.

Due to the present restricted apron area for run-ups, hard power cannot be pulled to get the oil temperature high enough to boil off the moisture through the breather, a byproduct of the internal combustion engine. Greg decided this year 'to increase the run-up rpm to 2,000rpm, that actually was fairly spectacular, and the crowds really loved that but then we noticed we were blowing over the stanchions (roped off crowd control area) behind the aircraft and have now reduced it to 1,800rpm for a minute or two.' Greg went on to say that 'the engines love to run at that rpm, the coolant temperatures and oil pressure come up to where they should be.'

The museum is investigating the possibility of a fifth spare engine to be set up as a 'Quick Engine Change' one to be kept on hand. Greg mentioned that the Merlin engine could easily and quickly be changed. Remember he said, 'this was a wartime engine and you did not have the time to be taking days for an engine change.' A full crew, everybody ready to go, and all the tools, the engine could be changed in two to three hours, four bolts and all the cannon plugs.

The museum's Avro Lancaster Mk.X, uses 100-octane low lead avgas fuel from the number two tank only and has been modified to use the source of the priming fuel from the same tank. The engines are Packard-built 224 Merlin engines that had been installed on the Canadian Mk.X Lancasters. Greg mentioned that 'the essential run-up crew in the cockpit is the pilot and the engineer, kneeling on the floor beside his panel as we took his bench seat out.' He also said 'I am around the aircraft all the time every day, so I let the other guys do the run-up most of the time. I will run it from the pilot's seat maybe twice a year, but the engineer's position more often.'

The engines require Aeroshell 120W oil and it is bought by the barrel for £700 ($1200). At the end of the season the engine oil screens are cleaned, 20 gallons (76 litres) are disposed of and recycled. Greg works within an annual budget to keep the Lancaster going as a major museum, Alberta, Canada, and world attraction. There are only four running Lancaster examples running worldwide. The budget, plus being able to trade extra/spare parts on hand for work and needed parts, in addition to Greg's ingenuity makes the whole thing possible. An example is Greg's battery for the Lancaster, it is from a Cessna Citation jet. The battery no longer met the specifications for the jet and was being discarded, but Greg realised it was so powerful that it could crank all four Merlins on the Lancaster, at once, and retrieved it for the museum!

LAHC
www.lincsaviation.co.uk
East Kirkby Airfield, Spilsby, Lincolnshire, England PE23 4DE
The privately run museum is well signposted on the A155 Sleaford Road with an imposing entrance featuring the original wartime entrance road and the guardhouse being replaced by a memorial to Nos. 57 and 630 RAF Squadrons who flew from the wartime airfield. Approaching the Admissions building the new large hangar, built on the previous T2 hangar site, catches your attention. The museum's whole programme is designed to recreate the wartime atmosphere of a Bomber Command RAF airfield. There is a Memorial Chapel containing the 848 names of personnel who gave their lives operating at East Kirkby. A recreated 'NAAFI' (Navy, Army and Air Force Institutes) provides further atmosphere and is a full-service restaurant. The sights, memorabilia, and sounds – recorded wartime music – greet you as soon as you open the door.

The Lancaster, if it is on the hangar apron, is visible through the windows close by. Off to the right is the restored Control Tower, also called Watch Tower, which has recreated sounds of R/T, Radio Telephone, transmissions and exhibits of WAAFs, Women's Auxiliary Air Force, manning the control desk. The original windsock is on display, which has a bullet hole in it, the result of an airfield attack by a Luftwaffe JU88 night intruder. The intended target was returning landing Lancasters from the 3/4 March 1945 Ladbergen raid; all landed safely, and then the airfield was raked with machine gun and cannon fire. Four personnel were severely injured and there was one fatality in the No. 57 Squadron briefing room.

'The Lancaster Experience' at East Kirkby is enjoyed in many ways – a visit to the museum and displays and, if the timing is right, witness the Lancaster taxiing, an interior aircraft tour, a taxi ride or the VIP taxi ride. The weekend schedule is three taxi rides available per day and the two-session VIP ride, held midweek, is often booked sixteen months ahead. The VIP ride is a full day package starting with a morning crew briefing and two runs morning and afternoon, divided into two groups you get to watch the other group on their taxi run. After a prepared lunch the groups switch positions. A visit to the cockpit and an external walk around completes the day. Crew members are on board to answer any questions.

There are six qualified taxi 'Pilots' available and two full-time Maintenance Engineers on staff. The operating crew consist of a Pilot, Flight Engineer and Safety Officer. The Safety Officer normally sits in the open doorway observing areas not visible to the crew in event of public ingress. The manoeuvring area is very confined – bad for the crew, good for the public as they can get really close to the taxiing Lancaster. The day before the taxi runs 'Just Jane' has the oil and coolant levels checked plus the tyre pressures.

On the day of the run the aircraft is towed out to the pad for a visual inspection at 0900. By 0915 the visitors are arriving and booking in. There is an assigned co-ordinator, 'Ops Officer', to look after the visitor group at all times. This is followed by a visitor briefing by the crew on Bomber Command and what the aircraft will do on the taxi ride and what the 'riders' will do on board. The Safety Officer will assign positions on the aircraft based on specific requests, 'my grandfather was a tail gunner,' and general preference. A normal visitor complement is nine filling the bomb aimer, behind the Pilot and Flight Engineer, Wireless Operator, Radar, Tail Gunner positions and three people standing in the mid upper turret.

The landing strip is checked suitable for taxiing, not too wet and soft and the Pilot checks the weather is suitable, and that the fire crew is in position. Andrew Panton, General Manager, Financial Officer and Lancaster Taxi Pilot, mentioned that 'the only other things that would cancel the taxi run would be strong winds and snow. The rain is not a problem except for visibility and that the ground crew get wet!' The longest straight-line taxi available is 1,936ft (590m) but that is part of a 30-acre (12-Hectare) field giving lots of room for manoeuvring. All taxiing is done in the tail down position.

Prior to start the aircraft battery is only used for lighting, radio and intercom and an external trolley accumulator is used for the heavy electrical loads of the engine start. The Lancaster's four engines are numbered left to right, one to four, as seen from the cockpit facing forward. It is interesting that the left and right side of the aircraft, facing forward, are named after the maritime designation as port and starboard. All aircraft have a steady red light on the left-wing tip and a steady green light on the right-wing tip, once again following the tradition of the seas. A steady white light is on the tail.

The sequence of the engine start, performed by the Flight Engineer, is number three engine, starboard inner, followed by number four, number two and number one. Once started, the engines are idled to check temperatures and pressures are in the normal range prior to taxiing. At the same time the flaps are raised, the bomb doors are closed, and the trolley accumulator is disconnected from the aircraft. The Pilot then gets a radio check with the fire crew to make sure of open communications. The aircraft is then taxied away from the hangar towards the grass strip. A bottleneck has to be carefully negotiated between an archive building and the public enclosure. Once again it is great for close public viewing of the Lancaster but a challenge for the Pilot.

Taxiing is done at 800–1000 engine RPM which is more than sufficient on the hard surface to keep moving. The pneumatic brakes are operated by a brake lever on the control column and differentially by applying rudder pressure – no toe brakes seventy years ago for Avro. 'Just Jane' is taxied at just over 50 per cent of its maximum wartime weight with 300 gallons (1364 litres) of Avgas on board. Fairly easy to make shallow turns on the hard surface, it does require use of the outboard engine on the grass strip.

The taxi procedure is fairly standard with a figure-of-eight taxi pattern followed by a run up into wind performed at the end of the grass strip. The engines are run up to 1800 to 2000 rpm, the brakes released, and the visitors are given a faster run, 30–40 knots, down the strip to give them a feel for the aircraft as it accelerates. The noise and the vibration take everyone back in history to the start of an operational sortie takeoff with crew survival unknown. Indeed, it is a sobering and emotional thought.

The Lancaster is then returned to the hangar apron for the shutdown procedure. The aircraft is guided by safety marshals as it passes close to the public enclosure. The engines are run up to 2000 rpm in pairs to clear them out after the long idling time, first the inboard pair and then the outboard pair. The bomb doors are opened followed by putting the flaps down. Then the four master fuel cocks are turned off simultaneously.

'Just Jane' is supported and maintained by the revenue generated by the taxi experience. The taxi experience started in 1995, limited to manoeuvres on the concrete area, to be followed eleven years later with the use of the grass area. Approximately 900 people annually enjoy the taxi experience. Sales and ops officer duties are performed by Fred

Panton's granddaughter, Louise Bush. The LAHC is working on plans to get 'Just Jane' airborne again. Judging by the museum's past performance that is a very attainable goal. Andrew Panton and Louise Bush form the only brother and sister Lancaster taxiing crew in the world, Louise is also a qualified Lancaster Flight Engineer! Some museum, some crew, some accomplishment – Fred would indeed be proud.

CWHM
www.warplane.com
9280 Airport Road, Mount Hope, Ontario, Canada L0R 1W0

The other three museums in this book are unique, with some similarities, but the CWHM is the only museum in the world, in 2018, where as a museum member you can purchase a Flight Certificate to have a one-hour flight in an Avro Lancaster. The Flight Certificates are valid for one year and, after the schedule is decided, bookings are made in April of each year. The Lancaster is not the only aircraft available for flights. Presently a total of twelve aircraft are/will be available for member flights. They range in size from the de Havilland Tiger Moth to North American Harvard to Beechcraft Expeditor to Douglas Dakota and North American Mitchell. There is also a Consolidated Canso flying boat. The flights in these aircraft are 20 minutes.

The flight status of each aircraft is displayed as they go through maintenance inspections or restoration to flying status. Some aircraft are available year-round and others, like the Lancaster, are available during the flying season only. The Flight Certificate works on the principle of an upgraded membership with a portion issued as a tax receipt because the CWHM is a registered charity. The reservations may be made online or by phone, some museum, some flight, some opportunity.

The mandate by the licensing authority, Museums Canada, is for the CWHM to collect, restore, and maintain, on ground or in the air, the aircraft collection. The thought being that not everyone can come to the museum to see the collection and that the museum should be able to take the aircraft to them. Don Schofield, former Chief Pilot, mentioned that in keeping with the mandate he had flown the Lancaster over every Canadian Province and nineteen of the forty-eight continental states of the USA.

The Museum will consider anything that generates revenue or a donation. This includes Airshows, Memorial Events, such as the Battle of Britain Day in September, Feature Movies, Photo Shoots and of course the members' rides. This means that it is possible for all Canadians to have an opportunity to see the Lancaster in the air within a reasonable distance of their home and not have to drive to Hamilton, Ontario, to see and hear the famous Merlin engines in flight. During August and September 2014, the population of the United Kingdom were able to enjoy the sights and sounds of the only two remaining airworthy Lancasters, the BBMF and the CWHM aircraft, flying in formation.

In its sixty-ninth year the Canadian Warplane Heritage Museum FM213 (KB726) Lancaster returned from the land of its birth to the land of its design and flew the same skies that many of its predecessors had flown in anger and in peace. Together with PA474, the Battle of Britain Memorial Flight Lancaster, it visited the old RAF Lancaster haunts and RCAF Stations, which once echoed to the unforgettable sound of four Merlin engines in harmony. In 2014 these two museums put on the best aviation remembrance show certainly of that decade and maybe longer. Some Lancaster veterans, and there are few now, said it would be the best show of the century with two Lancasters in formation flying out of a former Bomber Command Lincolnshire Station, Coningsby to all parts of the British Isles.

When asked about the serviceability of the Merlin engines with so many moving parts Captain Don Schofield, former museum Lancaster Chief Pilot, said tongue in cheek that 'as long as there are people with money flying Mustangs there will be no insurmountable problem with engine parts.' There is a lot of after-market remanufacturing, refurbishing, rebuilding and exchange of parts to keep 'em flying. All the CWHM museum aircraft are overly maintained because of the age and nature of the aircraft and the Lancaster in spite of its size is no exception.

The Merlin engine has a relatively short life compared to the engines of today, 400 operating hours to a major overhaul. A minor top overhaul costs roughly $120,000 to $150,000 (£67,000 to £83,000) and the supercharger failure in the 2014 UK tour eventually cost approximately $250,000 (£138,000) or so, as the debris had ended up in the crankcase. The Merlin is fine if you are established in cruise and droning along, that is what it was designed for, but if you have a lot of condition changes that is when the problems start to arise. A series of touch and go landings with many power changes is a good example of this.

I asked Schofield to describe a typical member's trip and he recounted the following information that I have turned into my narrative with occasional direct quotes. Away from base, the transport to the airport is arranged for a set time for the entire crew. Arrival at the airport is a well-rehearsed scenario after doing it so many times. If the aircraft is on static display the airshow organisers are instructed to move the spectators away from the aircraft 45 minutes prior to the flight and refuelling if necessary. The flight crew drop off their flight bags and take off for the airshow or a crew briefing. The pilots and flight engineer attend this briefing, which covers such items as abnormal, emergency and alternate operations as necessary due to aircraft, air traffic control and weather.

Returning to the aircraft the crew talk with spectators outside and inside the aircraft if it is open for tours. This can sometimes lead to very long days for the crew. The daily inspection is performed by the ground crew and any discrepancies as well as quantities added are entered in the log book. Reaching the engines does require the use of ladders or stands, which must now be removed. The pilots and the engineer(s) walk around the aircraft and do an external visual inspection together, starting at the right rear entrance door. Schofield said for spectator purposes 'the pilot takes the log book from the engineer which is then inspected ... for all checks and quantities added. The pilot in command, outside the aircraft, signs the bottom, shakes hands and passes the log book back before boarding the aircraft. Almost a mirror of what happened in WWII.'

Entering the aircraft, the pilot is constrained by the 'First prop(eller) turning (FPT).' The Lancaster is allocated a taxi or show time and using this, the time is worked backwards to the FPT. An airline style document called the Quick Reference Handbook is used for all sequential checklists such as 'On entering Aircraft' check, Flight Engineer's check, Pilot's check, Before Start check, After Start check, Before Takeoff check, After Takeoff check, In Range check, Before Landing check, After Landing check, and Shutdown check. It is a challenge and response checklist with the right seat pilot reading the checklist and the pilot in command, always in the left seat, responding with the correct responses.

Once the four engines are started, no two-engine taxi on departure taxi allowed, and the After-Start check complete, it is time to taxi to the runway. Keeping in mind, because of the nose high attitude, you cannot see straight ahead to taxi. Taxiing is done by visually maintaining the same distance from each side of the taxiway. The Second World War Avro Lancaster brake system is activated by a hand grip on the control column, the more you move the lever the more breaking you get, and rudder pedal displacement, which give you differential braking. The tail wheel is fully castoring and allows the tail to swing with differential braking. Power is increased to get the aircraft rolling and then differential braking and the use of the outboard engines, #1 or #4 as required, will initiate and stop a turn. The brakes are pneumatic and a light aircraft at idle power uses the brake pneumatic pressure faster than the air pumps can replenish. The only way to recharge the brake bottles is to stop the aircraft and bring up the power on #3 and #4 engines, the ones with the air compressors, and let them roar away to build up pressure again – quite confusing for some onlookers but now – you know!

Talking about hydraulics Schofield said, 'that is what really killed Pilot Officer Mynarski – the WWII hydraulic (system) was high volume and low pressure. Huge pipes, some up to an inch in diameter that used something called Girling fluid. This fluid was vegetable-based, highly toxic and, most importantly, very highly inflammable fluid.' The designers did not use hydraulics if at all possible and hence the pneumatic brakes. The high-volume hydraulic users were the landing gear and bomb bay doors.

The brakes, which are actually quite small, are all pneumatic. When you release the brake lever the air is vented overboard and Schofield mentioned 'that if you are in the airplane or close by outside you will hear the hissing, puffing and wheezing of the air going overboard. When the brakes get really hot the air also smells really bad,' and now – you know.

The main wheel tyres have old fashioned inner tubes on the Lancaster and are a huge 64 inches (163cms) in diameter. They are inflated to 54 psi (372kPa) and designed to work off grass. The museum spent approximately $80,000 (£42,000) on ten new tyres. To save the tyres the museum has a surface landing preference – grass, asphalt, normal concrete and to be totally avoided if possible, high traction concrete.

Prior to departure all the critical emergencies are rehearsed such as rejected takeoff and engine fire in the air. In my airline flying days it was referred to tongue in cheek as the 'I will, you will' procedures. Lining up on the runway Schofield commented, 'you will notice that the airplane just doesn't stop – it rolls forward a few feet. This is to straighten the tail wheel.' The brakes are held on by your hand and the inboard engines, #2 and #3, are run up to 30 inches, zero boost for the CWHM Lancaster. All the Lancaster engines rotate in the same direction and if the power is applied to all engines simultaneously you would not be able to control the aircraft from turning sharply and departing the runway.

Schofield then explained that, 'you take your thumb (right hand) and push up the #1 throttle and as it passes #2 and #3 (throttles) you catch them with the palm of your hand and then you extend your pinky and grab #4 so they are all going up (power) in unison. #1 is leading and #2 and #3 are following, followed by #4.' This counteracts the asymmetric power due to the direction of propeller rotation until the rudder becomes effective around 40 knots (74kmh). The pilot in the right seat then will put his left hand behind the throttles to make sure they do not retard due to vibration. When takeoff power is set the 'right seat pilot' taps the pilot in command's hand and takes over setting the throttles. The pilot in command can now put both hands on the control wheel. Any tendency to swing at this airspeed will take full application of rudder immediately to stop the swing.

Approaching 50 knots (93kmh) the tail starts to feel light and the control column is eased forward. This brings the aircraft into a level attitude on its main wheels and the cockpit actually goes down 4 feet (1.2 metres) because the aircraft is no longer in its tail down ground position. The aircraft takes off at 100 knots (185kmh). Once the aircraft is safely airborne Schofield mentioned that 'you put the brakes on. Those wheels weigh 780lbs (354kg) each and they are spinning, they have a torque factor all of their own.' When the wheels are stopped rotating the pilot in command signals for gear up, the right seat pilot from now on gets very busy.

As well as the museums and their staff there are other separate organisations that specialise in looking after the operational Merlin engines. As previously mentioned The museums often subcontract the specialised repairs, overhauls and restoration to companies throughout the world, one of these companies is Vintage V-12's of California.

Vintage V-12's

It is hard enough to pronounce but knowing where it is – impossible except for those interested in vintage aircraft piston engines. Amongst those engine aficionados, Tehachapi, California, is known world-wide and a very famous location. It is the home of Vintage V-12's, a vintage aircraft engine restorer and support services for the Merlin engine and other famous names such as Allison and Bristol. With a reputation to back up its acquired knowledge and skill of its mechanics, Vintage V-12's is a company supporting the Rolls-Royce and Packard Merlin engine. It is keeping the weekend warbirds, Reno racers and museum artefacts in the air, not to forget the ground running exhibits of the Merlin engine.

Forty years ago, Mike Nixon started his own shop in Chino, California, overhauling warbird engines after gaining experience working with his brother. In 1981 the name was changed to Vintage V-12's Inc. followed by a move five years later to Tehachapi. The workshop facility has over 10,000sq. ft. (929sq.m.) of office, machine shop, assembly area and small parts storage. To date the company has overhauled over 375 vintage engines and is adding to the total annually. The best companies are built through teamwork and Jose Flores is the Shop Foreman supported by a great team of mechanics.

Flores joined the company thirty-three years ago and started his in-house learning experience with menial maintenance tasks. His experience and knowledge grew so much during the years that he now owns and operates this very unique speciality company. Flores continues the in-house mechanical skills learning process and has accumulated a formidable team experienced in the restoration processes of the Merlin engine. Ninety-five per cent of all work is done by Vintage V-12's itself, thus maintaining full control of the restoration.

The engines arrive in all sorts of condition from all over the world. Vintage V-12's has restored engines in Russia, Belgium, Germany, Canada, Australia and South Africa as examples of its world-wide customers. Air freighted to Los Angeles the engines are trucked to Tehachapi. Some are in specially built wooden crates and some in the original World War II containers. When asked to comment on the engine teardown findings Flores said that, ' It depends on the amount of hours they have run but the blade rods are the most common part to wear as it is the last place to get oil. Our worst enemy is rust as over half of the engines out there don't run enough to keep them healthy.'

Vintage V-12's is in the business of engine restorations and use their parts for their own engine rebuilds. However, Flores indicated that he does appreciate what people are trying to do with vintage aircraft and the company will help out if it can. An example would be some seals and gaskets supplied for the Bomber Command Museum of Canada's Avro Lancaster. It does about fifteen to twenty engines per year for aircraft that have turned out to be an investment. It is the old story of supply and demand and the Merlin-powered aircraft are sought after, dare I say it is for the sound!

The stated goal of the company is 'to restore the customer's engine to meet or exceed the same level of performance, durability and reliability as the original engine.' The improved materials and technology have enhanced engine serviceability compared to those days of wartime struggle to get the engine built and into an airframe to fight the enemy. Vintage V-12's offers a full range of original parts and consumable parts such as gaskets and O-rings housed in a 7,000sq.ft. (650sq.m.) of warehouse space. The Bomber Command Museum of Canada and Canadian Warplane Heritage Museum have used their services to keep their Merlins serviceable.

The seven-step restoration process is an orderly set of steps that the engine follows for completion of the project.

1. Each project is assigned a work order number and a file is opened to record the condition of the parts, all measurements, repair work required and who accomplished the work.
2. The engine is completely disassembled, and a preliminary inspection is made to assess any damage, wear and the big enemy, corrosion. The parts and components are cleaned.
3. This step is the major inspection using Non-Destructive Testing to verify the integrity of critical parts prior to reinstallation. The testing includes Zyglo inspection for non-magnetic parts and Magnaflux inspection for steel parts. Parts are checked for compliance with manufacturer's specifications. The administrative process is to review all Service Bulletins and Technical Notes and ensure they are carried out for the particular engine. The applicable parts are listed for repair.
4. Repairs are accomplished onsite, maintaining manufacturer's specifications.
5. Accessories are kept in stock and ready for installation, items such as overhauled starters, coolant pumps, after-coolant pumps and ignition harnesses.
6. Engine assembly is conducted under frequent inspections to ensure quality control.
7. The final step in the process is for the engine to be run on a test stand for at least four hours to allow the critical components to break-in and also ensure that performance standards are met. A final inspection is carried out prior to the engine being shipped to the customer.

The Vintage V-12's Warranty states that 'our engine restorations are guaranteed for 50 hours or one year of operation.' The Merlin engine takes three to four months for a total restoration and very often the aircraft is kept flying during the restoration by having a spare engine also restored by Vintage V-12's. There is no doubt that Vintage V-12's is determined to keep the Packard and Rolls-Royce Merlin engine flying for many years to come. The hundredth anniversary of the engine's birth is only fifteen years away and their website banner says it all, 'Experience, Knowledge, Craftmanship'.

Every year in the high desert of Northern Nevada, USA, near the gambling and entertainment centre of Reno, the roar of Merlins can be heard as they try to outperform each other. Lake Tahoe and the Sierra Nevada mountains are close by the Stead Airport where, every September, the world-famous Reno Air Races are held. Their official name is the

National Championship Air Races for sporting aeroplanes in different size and performance categories.

The races are a multi-day event designed for the aviation enthusiast billed as 'the world's fastest motor sport' as it features multi-lap multi-aircraft races with extremely high-performance aircraft on a closed ovoid course marked out with pylons. The length of the course varies as per the speed classification of the contestants. The Unlimited Class record is 507mph (816kmh) and was set on a six-lap race around an 8½ mile, 13.7 kilometres, course in 2003. This was set by a North American P-51D Mustang, racing-tuned of course! The races were originally held at a dirt airstrip called Sky Ranch airfield before moving to their present location at the former Stead Air Force Base.

It is a week-long celebration of pure speed at low altitudes over the desert floor and includes stunt flying, civil airshow and military acts combined with a large civil and military static aircraft display. The star of the show has to be the British-designed and the British- and American-built engine named after a bird of prey, the Merlin.

In keeping with the title of this chapter there are individuals connected with the preservation of the Merlin, Martin Rouse, and the restored Merlin, Steve Hinton. These two stories are perfect bookends, 1945 and 2018.

Rouse's interest in aviation started in his youth and he particularly liked the Airfix de Havilland Mosquito model. The Mosquito also starred in the fictitious 1964 *633 Squadron* film complete with the sound of its Merlin engines. This interest always 'remained in the background' as he pursued his career in another industry.

Rouse's engine is a Rolls-Royce Merlin 25 number 150969/A439943. It was installed on a de Havilland XIX Mosquito MM677 RS-U of 157 RAF Squadron, RAF Station Swannington. It had been built at the de Havilland factory at Leavesden, Hertfordshire, and was installed in the aircraft with a Nash Kelvinator propeller. Flt/Sgt John Moore, pilot, and Sgt Trevor Westoby were flying the aircraft on the afternoon of 12 April 1945, practising night fighter tactics with Airborne Intercept equipment. At 1500hrs they were seen by the #2 aircraft in a dive below them. Subsequent investigation revealed structural failure in the port mainplane and the detachment of the port wing. The aircraft crashed between Feltwell and Southery, Norfolk, where the engine was recovered. The crew were both killed, they were both 22 years old and had been with 157 Squadron for just one week.

Thirty years ago, a friend told Rouse about his boss, at a classic car restoration business, who had a Lancaster engine and was restoring it. He thought that 'it was very interesting but way outside of my league' at the time. However, in 2017, an opportunity occurred to purchase a Merlin from Martin Ashby of Ashby Aero in Derbyshire. He went and had a look at it, thought about it, and is now the proud owner of a crash relic Merlin engine. He is from a mechanical background and had an interest in old cars and clocks, but it was 'always a dream' to own a Merlin and now it is reality.

The engine was in Ashby's workshop/warehouse among many aircraft parts and was not the only engine for sale. When asked why he chose this one Rouse indicated that 'it was a fairly complete engine compared to some of the others' but 'mainly because it was a Mosquito engine.' The engine was on a stand and had been washed down and given a coat of paint, 'it looked quite presentable, really.' Unfortunately, to date he has not been able to trace the people who recovered it. Ashby's had given him some basic information, which he has since added to with his own research including a visit to the National Archives at Kew, London.

To prevent further corrosion presently he is carefully removing as many parts as possible whilst trying to avoid further damaging the engine to remove the embedded Norfolk mud which is holding moisture inside the engine. Already some 'crystals of alum' have been discovered growing in the engine. It is suspected that the original RAF recovery team decided it was too much trouble to dig the engine out and just buried it. He is hoping to put it back together again with as many of the original parts as possible. In some places in the engine the Norfolk mud had set like concrete and it would do too much damage to dismantle further. Asked if he was going to run the engine Rouse said that 'you would have to replace so many parts that the engine would not be the original engine anymore.' The aim is that the engine will be a static exhibit, an historic artefact, possibly mounted on a trailer with accompanying information on the engine, the Mosquito and the crew and displayed at airshows etc. It will be a 'sympathetic restoration' said Rouse 'so as not to take away any history'.

At the other end of the engine spectrum is the highly tuned racing Merlin engine in Steve Hinton's North American P-51 Mustang. The Merlin variants were built to improve their power output for the bomber to carry heavier bomb loads to a higher altitude and for the fighter to fly faster at lower or higher altitudes as the aircraft variant was designed for. Hinton's Mustang certainly needs that power as he flies the turns of a measured course in air races or to achieve new speed records for a piston-engine aircraft. The fact that the modified Mustang, and the specially prepared Merlin engine, were chosen for these flights speaks highly of both their pedigrees, as we hope has been shown in the preceding chapters.

Steve Hinton is the son of the Steve Hinton who established himself in the air racing and air speed record books in the late 1970s and the 1980s. The elder Hinton has held a world speed record for ten years, won national championships and unlimited-class races. His record was established with a modified P-51 Mustang, named the *Red Baron*, on 14 August 1979. Subsequently he became President of the Planes of Fame Air Museum in Chino, California. He is a charter member of the Motion Picture Pilots Association. With that family history was there any doubt that the younger Steve Hinton would become involved with aviation in general and air racing specifically?

Working in the family business to gain aviation experience helped in his understanding of flight and on his sixteenth birthday he soloed a Cessna 150 and joined the ranks of qualified pilots. Building hours quickly during those four years, he flew heavier and more complicated aeroplanes, the Luscombe, Stinson L-5, Stearman, North American T-6 Texan and finally by age nineteen the North American P-51 Mustang. Air races and speed were never far from his thoughts.

In 2007 he was very fortunate to continue his 'education' and spent the summer at Vintage V-12's under supervision dismantling oil screens, cleaning the parts and assembling them again. Also that year, as part of a crew, he worked on the valve train and its timing. In 2008, after taking the required 'rookie school' course, he raced his first Mustang called 'Sparky' taking the trophy in the bronze category.

The next four years he raced the Merlin engine Mustang, named 'Strega', which normally gave at 60 inches of manifold pressure 3,000 revs per minute and 1,500 horsepower, but which in race power the boost was around 130 inches giving 3,400 revs per minute and approaching 3,400 horsepower. Depending on conditions this resulted in speeds around 500mph, qualifying for the gold category. This speed does not come easily. About 100 hours of maintenance is done for every flying hour and after twenty flying hours, the engine must be stripped down. It looks very much the worse for wear and is expensive to restore.

He has won the Unlimited Gold title at the National Championship Air Races seven times and in 2011 set a personal best of 499mph (803kmh) during a heat qualifying run. In 2013 he had the opportunity to change and to start flying the Mustang called 'Voodoo'. The scene was set for an attempt on a world speed record for piston-engine aircraft. This is where

Bob Button and Joe Clark enter the picture and become very important, along with a supporting cast of aviation experts.

The aircraft for the 2017 attempt on the record was a North American P-51D-25-NA (original serial number 44-73415) built in 1944 at the North American Aviation plant in Inglewood, California. In 1951 it was transferred to the RCAF as a Mustang IV and served in Fighter and Tactical Command. Sold off in the late 1950s it returned to the USA and had an extensive chequered history until bought by Bob Button of Bob Button Transportation Inc., in Wellington, Nevada. Button teamed up with Joe Clark, CEO of Aviation Partners Inc., a world leader in innovative aerodynamics including winglet technology and sponsor of the challenge. That is the background of the Mustang called 'Voodoo', but what about the Merlin engine?

The engine target was for the Merlin to give 140 inches of boost resulting in 3,400 horsepower. Unfortunately, 'oil issues and carburation problems gave only 121 inches of boost,' according to Hinton. The whole engine package was a calculated endeavour to get the best blower ratios to combine with the best components. The top end of the engine was the civilian, or 'transport Merlin' as Hilton referred to it, 600 series engines and the lower end the 724 engines. These engines were used in the Canadair DC-4M, North Star and C-4 Argonaut. The core was designed to withstand long-range cruising conditions of heat, stress and vibration.

The supercharger housing was from the Packard V-1650-9 series and the actual gearing from the V-1650-23. The sweep of the supercharger turbine rotor was adopted from the P-82 Twin Mustang with a more efficient 40-degree cut. The propeller gearing ratio was changed to .420:1 giving a slower rotation. The pistons, aftermarket products, were changed to 5.5:1 instead of 6.0:1 engine compression ratio. Camshaft overlaps were adjusted for the new specifications of the engine. Connecting rods were beefed up with a change to Allison G-6s to link the pistons to the crankshaft.

An added feature to the record attempt engine was the anti-detonation injection device. Installed in place of the after-cooler it injects a 50/50 mix of distilled water and ethanol to cool the compressed intake air prior to entering the combustion chamber. Finally, to handle the greatly increased horsepower the crankcase had bracing, and a spine installed to stiffen the core of the engine. It certainly was a well thought-out and engineered Merlin engine for the record attempt.

The Fédération Aéronautique Internationale (FAI) is the governing world-wide organisation of aviation sports including the verification and holder of all data. The speed attempt was to better the speed of the retired record of 528mph by Lyle Shelton's Grumman F8F-2 Bearcat *Rare Bear* in 1989. Because this is a retired record it can never officially be beaten. On 2 September 2017, the Merlin-powered P-51D Mustang 'Voodoo' with Steve Hilton at the controls made four passes over a 3km (1.86m) course at an airport at Challis, Idaho, achieving an average speed of 531.53mph (854.1kmh). This qualified as a verified world's fastest speed of a propeller-driven piston engine aircraft in the Class 1-e category; well done Merlin, well done Steve Hinton and team!

Conclusion

Sixty-two years ago, as my 'airspotting' friend Bill Powderly and I stood on the cold and windswept observation level of Dublin Airport, Ireland, looking at the landing Canadair C4 Argonaut, Bill turned to me and said, 'Looks like a DC4, sounds like a Lancaster. Can you imagine what a bomber raid must have sounded like?' There you have it, notwithstanding the performance aspects of the Merlin, which the pilots enjoyed, most people remember it for its unmistakable sound. Another 'airspotting friend', John Kimberley, was at a boarding school and the evening 'prep' time for homework coincided with the evening Dan Air Avro York flight from Dublin Airport. Hearing the distant sound of the Merlins, John would raise his hand and ask to go to the washroom. He would incur a penalty, 'writing lines', for not going to the washroom before 'prep' started. He would run out to the school yard and see the flashing anti-collision lights of the York but, more importantly, he would enjoy listening to the sound of the four Merlins in the evening sky! Such was, and is today, the allure of the Merlin.

The Merlin story started with a man who was dissatisfied with his purchased car. Henry Royce was the man. He was a successful electrical designer and manufacturer, and the Decauville was the car to be improved. In 1904 Royce had his own improved version on the road and the famous Rolls-Royce smoother ride had its first incarnation. This would soon come to the attention of Charles Rolls, a sporting entrepreneur and keen advocate of the motoring industry. The genius engineer Royce and the larger-than-life promoter Rolls bonded. So, began the design and development of bigger and better automobiles, the Rolls-Royce.

Unfortunately, it would be his undoing in the end. Rolls was not content with the automobile but embarked on a ballooning adventure followed closely by learning to fly aircraft. Rolls was killed during a flying competition in 1910 and so ended the personal Rolls-Royce partnership. He had just become, months earlier, the first pilot to fly a return trip across the English Channel. Although the ten-year partnership had been relatively short the momentum established by Royce and Rolls would propel the company to great achievements in the coming years. The first Rolls-Royce aero engine, the Eagle, appeared in 1915. This engine was a great asset to the Allies during the war and subsequently powered the Vickers Vimy on the first Transatlantic crossing by air.

Royce continued to develop both the Rolls-Royce automobiles and the aero engines during the inter-war years. Royce drove himself mercilessly to achieve engineering excellence in both industries with the result that he passed away in 1933, some say from ill health caused by chronic exhaustion. Once again, the Rolls-Royce momentum carried on his engineering conception of a liquid-cooled V12 engine, the PV12. Two years later the Merlin engine flew for the first time in a Hawker Hart. It is a tragic irony of the Merlin story that Royce never saw his Merlin engine go airborne and that Mitchell never saw his Spitfire go into full production. Their posthumous contribution to the Allied war victory would go down in history.

However, of the two-single engine monoplane pre-war fighters it was not the Spitfire that went airborne first, it was the Hurricane. The Merlin had been successfully installed in its fighter fuselage and in June 1936 the first government orders were on the Hawker books. Previously in March 1936 the Spitfire had taken to the skies. In three years' time, 1939, the backbone of RAF Fighter Command would be these two aircraft; one easily built and repaired, the Hurricane, and the other a sophisticated labour-intensive design, the Spitfire. The commonality of the Merlin engine and its variants would support these two great airframes to develop to a high level of performance and reliability. They would both shortly be called upon to defend, very nearly to their own total demise, the country of their birth, Great Britain.

However, first the Merlin would deploy to France in the Fairey Battles and then the Hurricane. A losing land battle finally relegated the Merlin to the skies above the retreating army culminating in the evacuation from French soil at Dunkirk in June 1940. The battle shifted to the attack on Great Britain and was very similar to the Battle of Waterloo in that the Merlin-powered Hurricanes and Spitfires narrowly won – 'The nearest run thing you ever saw in your life.' The fighters were aided by the radar defence system in the struggle against the overwhelming odds. Of course, it goes without saying that the machines could not do this alone but that it took the dedication and bravery of the men and women involved to achieve success.

To supplement the pure fighter a unique Merlin-powered wooden aircraft, the Mosquito, was developed. Extremely fast and versatile, from fighter to bomber and everything in between, this remarkable aircraft filled many roles. It participated in long-range bombing missions to Berlin; pinpoint bombing raids such as finding and destroying Gestapo Headquarters; high and low-level photo-reconnaissance missions; fighter escort duties and marking targets for the bomber force. The 'Mossie' did it all. The Mosquito was a very successful companion to the Hurricane and Spitfire and proved that the Merlin engine and the British airframes were at least equal to the enemy aircraft. The difference would be up to the crews.

A novella written by Frederick Forsyth, *The Shepherd*, catches some of the magic of the Mossie. It concerns a de Havilland Vampire pilot returning to England from Germany in 1957 with complete electrical failure and lost in the fog. A mysterious de Havilland Mosquito appears and shepherds him home to safety. The Vampire pilot later learns his saviour is Johnny Kavanagh, a pilot who had disappeared over the North Sea on Christmas Eve 1943, 14 years earlier. The story has often been read on the radio during December as an unusual Christmas ghost story.

With the immediate threat of invasion stopped in 1941 and it was the time to take the battle to the enemy. Now was the time for the Merlin and the medium and heavy bombers to wreak havoc on the manufacturing, transportation and infrastructure of the enemy. Many aircraft were using the Merlin and through careful planning, foresight and effort, by both Rolls-Royce and the government agencies, the supply of engines continued to support the world-wide advance of the Allied forces.

When America joined the Allied forces, it brought with it an American fighter, the Mustang. Initially it was a mediocre performer but the addition of the Merlin engine turned it, by the latter stages of the war, into the best all-round fighter. With internal and external fuel tanks, it could now escort the bombers to their destination and back, plus pick individual targets of opportunity deep in enemy-held territory. The Merlin, six years after its first flight, was still proving its worth following Royce's courageous decision to design a liquid-cooled engine.

Until the end of the war Rolls-Royce continued to develop the Merlin for a variety of applications. Each improvement increased the power of the engine and also tailored it to specific altitude, environment and aircraft manufacturer applications. Improvements included the number of stages and speeds of the supercharger. The Fleet Air Arm requirements were far different from the high-altitude fighter. Shadow factories were established away from the main areas of battle to maintain the supply of Merlins. This allowed a continuous supply of engines to the operational squadrons. Packard in America could be termed such a factory, as it produced Merlins under licence and shipped them to Britain already installed in the Canadian Hurricane, Mosquito and Lancaster, as well as the American Mustang. This was in addition to the engines shipped separately from fuselages. The Rolls-Royce Hillington, Glasgow, factory also repaired Merlins and modified the Packard engines to bring them up to the latest modification level.

The engine, and the bird, can still be heard above the hedgerows of England eighty-five years after F. H. (Henry) Royce came up with the basic concept of the Merlin. The RAF, Battle of Britain Memorial Flight, the Canadian Warplane Heritage Museum, and other museums and private individuals fund the restoration of the Merlin engine and their aircraft to flying status. Many individuals and organisations preserve and restore for display an example of the Merlin engine, some running, some not.

Why is the Merlin engine so revered among historians and aviation aficionados? Because it is loved for what it did, it allowed me and millions of others to grow up in freedom in post-war Europe, for what it is, a brilliant V12 water-cooled engine, and what it will be – a great engine which will keep its unmistakable sound for many years to come and will remind listeners of what happened 1939–45. The justification for the subtitle, 'The Engine that Won the Second World War', is the fact that the tide of war turned during the Battle of Britain in 1940 from the defensive, won by mostly Merlin-powered fighter aircraft, to the offensive, which included many Merlin-powered fighter and bomber aircraft. Without this turn of the tide Great Britain would have been invaded and the battle, and the war, would certainly have been lost.

Rolls-Royce Post-Merlin to Trent

Rolls-Royce Peregrine

Rolls-Royce continued piston engine development and production during and after the Merlin era. The Peregrine was the ultimate development of the Kestrel engine although it had the added benefit of six years' of Merlin development. The basic Kestrel engine was strengthened to allow greater supercharger boost pressure to be used. This gave a better power-to-weight ratio than the Kestrel and Rolls-Royce felt that this engine would be the fighter standard for the impending hostilities. A unique design feature was that the engine was available in left- and right-hand tractor versions. This would provide a balanced counter-rotating propeller installation for twin-engine aircraft, a common configuration for German aircraft but not for Allied aircraft. By 1943 the Peregrine engine had been cancelled, a victim of the successful development of the Merlin. It had been installed and saw service in the Westland Whirlwind and the second prototype Gloster F9/37.

Engine	Peregrine I
Horsepower (hp)	885
Number of pistons	12
Engine Capacity	21.2 litres
Bore	5.0 in (127 mm)
Stroke	5.5 in (140 mm)
Basic Dry Weight	1,140 lb (517 kg)

Rolls-Royce Griffon

The RNAS merged with the RFC in April 1918 to form the RAF. On 1 April 1924, the Fleet Air Arm (FAA) of the RAF was formed encompassing those RAF units that normally operated on aircraft carriers and fighting ships. Development of the Griffon engine started for the FAA in 1938 specifically to give high power settings at low altitude, to be reliable as it would be used in the single-engine carrier-based aircraft such as the Fairey Firefly fighter and anti-submarine aircraft. It was also adapted for use in the Supermarine Spitfire. With a very similar frontal area, about 5 per cent greater than the Merlin, it had a greater cylinder swept area and increased the bore by 0.6 in (15.3mm) to 6.0 in (167.6 mm). Engineering innovations continued the Rolls-Royce tradition, such as reducing torsional drive in the camshafts by taking the drive from the front of the engine in the propeller reduction gear and lubrication of the main bearings and big ends using a hollow crankshaft to provide more even distribution of oil. However, the Griffon was under the shadow of the Merlin and had a temporary halt to production in 1940 as resources were focused on the more successful Merlin. It was built in many versions of number of stages and different speeds of supercharging to give maximum performance at different altitudes. The Griffon rotated in the opposite direction to the Merlin, which presented a challenge to pilots due to the opposite torque as they converted from one Spitfire Mark to another. To alleviate the torque problem, the engine was also built with contra-rotating propellers. The Griffon was made until 1955 and can still be heard flying with the Battle of Britain Memorial Flight and at the Reno Air Races in several North American Mustangs.

Engine	Griffon 65
Horsepower (hp)	2220
Number of pistons	12
Engine Capacity	36.7 litres
Bore	6.0 in (167.6 mm)
Stroke	6.6 in (167.6 mm)
Basic Dry Weight	1,980 lb (900 kg)

Rolls-Royce Exe

The Exe was an experimental air-cooled piston engine that broke with the line of Rolls-Royce liquid-cooled engines and also the naming convention for piston engines. It was named after a river rather than the usual birds of prey. Rolls-Royce would later transfer the use of river names to its line of gas turbine engines. It was test-flown in a Fairey Battle in November 1938 but succumbed to the demands of the Merlin and Griffon engines and ceased production shortly after hostilities started. It was unique in that it had sleeve valves and pressure air cooling. Its twenty-four cylinders were arranged in X form of four banks of six cylinders each.

Engine	Exe
Horsepower (hp)	1,150
Number of pistons	24
Engine Capacity	22.1 litres
Bore	4.22 in (107.3 mm)
Stroke	4.0 in (101.6 mm)
Basic Dry Weight	1,530 lb (694 kg)

Rolls-Royce Crecy
Two stroke supercharged sleeve valve engine. Problems with piston cooling, terminated.

Engine	Crecy
Horsepower (hp)	2,729
Number of pistons	12
Engine Capacity	26.0 litres
Bore	5.1 in (129.5mm)
Stroke	6.5 in (165.1mm)
Basic Dry Weight	1,900 lb (862 kg)

Rolls-Royce Pennine
The Pennine was a prototype engine based on an enlarged version of the Exe. Once again it was an X shaped pressure air-cooled sleeve valve engine named in this case after the Pennine mountain range to signify that the engine was for commercial aviation. It was built in 1945 but never flew due to the advent of the turbo jet engine.

Engine	Pennine
Horsepower (hp)	2,740
Number of pistons	24
Engine Capacity	45.73 litres
Bore	5.4 in (137 mm)
Stroke	5.0 in (127 mm)
Basic Dry Weight	2,850 lb (1,293 kg)

Rolls-Royce Vulture

In the same X configuration as the Exe engine, the Vulture can be thought as four Kestrel cylinder banks mounted on a single crankcase driving a single common crankshaft. The engine development was suspended during the 1940 Battle of Britain as Rolls-Royce concentrated all its efforts on the Merlin. This probably contributed to its ultimate demise. Its reliability was poor when it entered service on the Avro Manchester due to lack of lubrication that caused connecting rod big end failures, unreliability due to connecting four connecting rods to a complicated arrangement with the big end bearing, and problems with heat dissipation from the twenty-four cylinders. The Vulture employed light alloy cylinder blocks, head and coolant jackets and steel wet cylinder liners. Seven bearings supported the crankshaft and the engine had a Skinner Union (SU) twin choke carburettor with two speed single stage supercharger The Merlin was already superseding the Vulture in horsepower, so a decision was made to substitute four Merlins for the two Vultures on the Manchester. This configuration became the Avro Manchester Mark III, which subsequently became the Avro Lancaster, Britain's leading heavy bomber.

Engine	Vulture V
Horsepower (hp)	1,780
Number of pistons	24
Engine Capacity	42.47 litres
Bore	5.0 in (127 mm)
Stroke	5.5 in (139.7 mm)

Rolls-Royce Eagle (the second)

Nearly thirty years after the first Eagle engine became airborne another Eagle first ran as the 1944 version. This would be the last in the line of illustrious Rolls-Royce reciprocating liquid-cooled engines. Once again, the need was foreseen for greater power from less weight. Rolls-Royce experience limits were now imposing a bore of 6 inches, due to aspiration and flame travel, and twelve pistons per crankshaft. Thus, a twin crankshaft in a flat H formation resulted for the new Eagle. A blade and fork connecting rod attachment system was used. Internal bracing and double-walled parts of the crankcase ensured rigidity. A two-stage supercharger and intercooler were standard. An innovation was three inlet ports per cylinder and two exhaust ports to a common pair of ejector stacks. The Eagle 22 was designed for contra-rotating propellers and it was installed in the Westland Wyvern prototype for carrier-based operations. The Eagle was never used in front line fighters as by then the line of Rolls-Royce turbojet engines was changing the face of aero engines.

Engine	Eagle 22
Horsepower (hp)	3,200
Number of pistons	24
Engine Capacity	46 litres
Bore	5.4 in (137 mm)
Stroke	5.125 in (130 mm)
Basic Dry Weight	3,900 lb (1,769 kg)

The engineering and scientific knowledge gained in the development and production of the first Eagle engine in 1915 to the last Eagle engine in 1944 was immense. The intervening years allowed Rolls-Royce to push the boundaries in all aspects of engine development. This included the actual hardware of the engine and the understanding of the control and movements of air during the intake, compression, combustion and exhaust phases of engine operation. The supercharger was a very important innovation of engine development as it achieved maximum power from the engine. This experience with the handling of air in the four phases of engine operation would help the Rolls-Royce engineers in Rolls-Royce's transition to the jet age. Rolls-Royce was not the first company, or Rolls-Royce engineering personnel the first designers, to think of other ways of developing power by changing the standard piston internal combustion engine to a new configuration, namely the jet engine.

One definition of turbine, a word with Latin and French roots, is a machine for producing continuous power in which a wheel or rotor, typically fitted with vanes, is made to revolve by a fast-moving flow of water, steam, gas, air or other fluid. Engineers would refer to this as the kinetic energy of a moving fluid converted into mechanical energy by the rotation of the bladed rotor. Two examples, which have been around for many years, are the wind-powered Dutch windmills for pumping water and the water-powered English mill used for grinding grain. The jet engine, then, is an internal combustion engine in which the expanding gases from the combustion chambers drive a turbine, which in turn drives a compressor to compress the air for combustion. Power is either from the expanding exhaust gases or taken mechanically (torque) from the turbine. A supercharger and turbocharger do the same thing, increase the pressure or density of the air to the engine, the difference being that a supercharger is mechanically connected to the engine and that a turbocharger uses its source of power from a turbine in the exhaust.

Not discounting the ancient times of Chinese firecrackers and Greek inventions, it was John Barber who, in 1791, received a British patent for *A Method of Rising Inflammable Air for the Purposes of Producing Motion and Facilitating Metallurgical Operations* and in it he described a turbine. Nearly one hundred years later Charles Parsons patented the steam engine and in accompanying documents clearly described a gas turbine. While Royce ran his ten-horsepower internal combustion piston auto-engine in 1903, Aegidius Elling, a Norwegian researcher and inventor, built the first gas turbine that could produce more power than was needed to run its own components and thus became the 'father of the gas turbine'.

Eighteen years later in 1921 Maxime Gauillaume patented the axial flow turbine engine. It used multiple compressor and turbine stages, which is very like present day jet engines. In 1926 Dr Alan Griffith, who would later join Rolls-Royce in 1939, published a paper titled *An Aerodynamic Theory of Engine Design*, which revolutionised the thinking that the turbine engine was very inefficient and would never be suitable for aviation. Dr Griffith, during his research at the RAE, designed an axial flow jet engine called the Metrovick F.2 that first ran in 1941. Metropolitan Vickers was the manufacturer and the engine development had been carried on by his partner Hayne Constant. An F.2/40 was installed in 1943 in a Gloster Meteor and flew as a possible engine contender. He proposed that an aerodynamic shaped blade would correct the situation of the flat blades 'flying stalled' and would dramatically improve their efficiency. Dr. Griffith went on to describe an axial compressor and two-stage turbine engine arrangement with a power take-off-shaft, which would be used to power a propeller. This early design proposal was a premonition of the turboprop engine to follow. A working model was later constructed and run by the Aeronautical Research Committee (ARC). Griffith's theory did work.

It is fascinating to think that ten years after squadrons of biplane fighter aircraft with rudimentary piston engines were duelling among observation balloons over the First World War trenches that inventors, researchers and engineers were contemplating a new form of aircraft propulsion, the turbine engine. One of these was Frank Whittle, later to become Air Commodore Sir Frank Whittle, OM, KBE, CB, FRS, FRAeS. Whittle came by his practical engineering skills honestly, working in his father's business, Leamington Valve and Piston Ring Company. His natural curiosity made him a constant reference library reader, especially in the fields of astronomy, engineering and the theory of flight.

He initially started with the RAF at the No.2 School of Technical Training at RAF Cranwell. His abilities, mathematical genius among other things, gained him a place on the Officer Training course at RAF College Cranwell. His training included flying training on the Avro 504 and he went solo in 1927, a childhood dream come true. Part of his course was that Whittle had to write a thesis to graduate. His was on the *Future Developments in Aircraft Design* and particularly the theory that a motorjet, a conventional piston engine that used compressed air to add thrust to the motor, at high altitude would be more efficient than a conventional power plant. He excelled at his studies and graduated in 1928 with a 'good, bad and ugly' record! The good was the Andy Fellowes Memorial Prize for Aeronautical Sciences and being described as an 'exceptional to above average' pilot. The bad was numerous warnings about showboating and overconfidence. The ugly was a dangerous flying incident that disqualified him from the end-of-term flying competition. Whittle continued to think about his thesis and eventually came to the 'eureka moment' conclusion of substituting a turbine engine for a piston engine and instead of a piston engine providing the compressed air for the burner that a turbine mounted in the exhaust could do the same thing.

Initial posting saw him continue his adventuresome ways and he was soon posted to the Central Flying School as a flying instructor at RAF Wittering, Cambridgeshire. Whittle showed his new engine concept around the base, which eventually came to the attention of the base commander. By late 1929 he sent his new project to the Air Ministry for consideration. The Air Ministry sent it on for further examination to Dr. Griffith, who concluded that Whittle's simple design could never be developed as a practical engine. He pointed out an error in Whittle's calculations and concluded that using a jet for power would be rather inefficient. A fellow pilot, Pat Johnson, was convinced of the design validity and had Whittle take out and file a patent in 1930. It was a cleverly worded patent in that it described an engine with two axial compressor stages and one centrifugal stage, he definitely 'hedged his bets' for future development. The fact that the RAF did not declare the concept a secret meant that Whittle retained the rights to the idea. Whittle met with British Thomson-Houston (BTH) but the company did not want to spend the necessary development money, so the idea withered on the vine. Just think of how history, namely the Second World War, would have been altered with Britain having a jet engine in the early 1930s and the subsequent development until 1939.

Whittle continued his RAF career and in 1932 started the Officers' Engineering Course at RAF Henlow, Bedfordshire. He graduated six months early with an aggregate score of 98 per cent in all subjects. This allowed him to take a two-year engineering course at Cambridge University and he graduated in 1936 with a First in the Mechanical Sciences Tripos. While at Cambridge he was a typical student, genius rather, with limited funds and did not/could not renew his patent for the jet engine. Two retired RAF officers, one he had been with at RAF Cranwell and who realised Whittle's position, proposed a partnership to act on Whittle's behalf and get financing for the engine project. Through family connections and independent verification of Whittle's plan's feasibility, the Bramson Report, the trio achieved financing through the O. T. Falk & Partners (Falk) investment bank. In November 1935, five years on from a possible first start, the jet engine could finally become a reality.

On 27 January 1936, a 'Four Party Agreement' created Power Jets Limited, which was incorporated in March 1936. The four parties were Whittle, Falk, the President of the Air Council and the two original partners together, Rolf Dudley-Williams and James Tinling. Whittle was placed on the Special Duty List to continue his work because he was still a full-time RAF serving officer. Power Jets entered into an agreement with British Thomson-Houston (BTH) in Rugby, Warwickshire, as the manufacturer of the prototype engine. By the end of the year the design was finalised, and parts were being completed for assembly. A further study of Whittle's design by Griffith was requested by the Air Ministry and resulted in a slightly more favourable evaluation, not surprising as Griffith had already started construction of his own turbine engine. The ARC decided to fund Griffith instead.

Notwithstanding the Air Ministry indifference, bias, or whatever, now Falk indicated they were restricting their future funding to less than a third of the original amount. Despite these set-backs the first Whittle engine, the W.U. (Whittle Unit), ran successfully on 12 April 1937. The prototype had combustion problems and the compressor was quite inefficient, but it ran. Whittle subsequently made a discovery about the vortex flow on the turbine, which would influence future designs. The ARC was now suitably impressed and convinced the Air Ministry to provide further funding although it would not be available for a year, more unnecessary delay. In spite of financial woes, the project continued. The W.U. did have some engineering challenges to contend with, one being the tendency to race out of control. Because of the dangerous nature of the work, the project was moved to a less active foundry location at Lutterworth, Leicestershire. Work continued there and when war broke out in September 1939, Power Jets had only a small workforce of ten personnel. Griffiths' work at RAE and Metropolitan-Vickers was of comparable size.

The stress of the engine development had taken a terrible toll on his health and in this regard his situation was similar to Royce's. Another visit to Power Jets in June 1939 saved the project from oblivion, the funds were running out again. This time the engine demonstrated a full power operation, the rpm had increased in two years to eight times the original rpm, for 20 minutes, and the Air Ministry was finally convinced, provided funds and ordered a flyable version. The problem now was to turn the W.U. ground engine in to one that could fit and perform in an aircraft; the result was the Whittle Supercharger Type W.1.

The engine was a reverse flow design with air from the compressor fed rearwards to the combustion chambers, then actual combustion in a forward direction with the exhaust flow rearwards once again, a figure 'S' on its side, if that helps visualisation. This design reduced the length, and hence the weight, of the engine. This same design principle was used in the very successful Canadian Pratt & Whitney PT6 engine twenty years later, and still is, in 2018, which powered more than fifty different manufacturers of aircraft and helicopters. The Air Ministry placed an order with the Gloster Aircraft Company to provide a simple aircraft to test the W.1, it was known as the Gloster E.28/39, and with Power Jets to develop a larger engine, the W.2. The W.2Y was also on the design table as a straight-through flow layout engine. A contract was issued for up to 3,000 W.2 engines per month in 1942. Rover was the only company to sign.

In 1940 Stanley Hooker of Rolls-Royce had met Whittle and they both were dealing with the same engineering challenges, the compression and resulted behaviour of air. Hooker oversaw the supercharging division of Rolls-Royce, which took the piston engine to new power levels. He later introduced Whittle to Hives, Chairman of Rolls-Royce, who agreed to have the company supply parts such as turbine blades and gear cases as a secondary manufacturer for the Whittle engine to help the project along. This showed the level of frustration of Whittle with Rover, that he would have an outside company, in effect a subcontractor, provide parts for his engine. This informal agreement would be an important catalyst in the history of Rolls-Royce as it faced its own transition to the jet age.

The airframe was ready long before the production engine, so Whittle assembled a one-off version called the W.1X – X for experimental – which ran on 14 December 1940. The stress of this endeavour caused Whittle to have a mental breakdown and he had to take time off. The exigencies and pressures of war greatly added to the normal development progression of peace time. The engine was installed in a Gloster E.28/39 (G.40) and during taxi testing, did some short airborne hops. Finally, on 15 May 1941, a prototype W.1 of 850 lbf (3.8kN) took to the air in the E.28/39 from RAF Cranwell for 17 minutes. It achieved a speed of 340mph (545kmh). Power Jets Ltd had done it, out of the ashes of financial disaster and Whittle's despair, the Phoenix rose. It was called simply the W.1. Within a short time, the engine powered the E.28/39 to 370mph (600kmh) and 25,000 ft (7,600 m). The world's best piston engine, the Rolls-Royce Merlin, was already being outperformed by this new engine, an innovation, and the historic transition had begun from piston to jet, and turboprop, engines. Rolls-Royce understood this and formed a Gas Turbine Collaboration Committee during the summer of 1941.

Rover created a laboratory and production factory at Barnoldswick, Lancashire, to produce the Whittle engine and future developments. The Whittle and Rover working partnership did not work out. Whittle felt that Rover could not deliver production quality parts for his engine and perhaps more aggravating was that he thought Rover was tampering with his design and the altered engine would be claimed as their own, not a healthy engineering environment. In nearby Clitheroe, Rover engineers, under Adrian Lombard, worked on a straight-through design dispensing with Whittle's reverse flow. The engines were designated the W.2B/23 for Whittle and the W.2B/26 for Rover. Sometime in late 1942 or early 1943 Spencer Wilks of Rover and Ernest Hives of Rolls-Royce met, with the agreement of their boards and the Minister of Aircraft Production, and decided to exchange production facilities and focus for the benefit of both companies. Rolls-Royce would assume the Rover jet factory in Barnoldswick and Rover would assume the Rolls-Royce Meteor tank engine in Nottingham. Rover maybe won in the immediate short-term, but Rolls-Royce certainly won the future rewards. Hooker found himself now as the chief engineer of the new factory manufacturing the W.2. At this time, a Whittle engine was installed in the tail of a Vickers Wellington bomber and air-tested at the Rolls-Royce Hucknall airfield.

Rolls-Royce assumed ownership of thirty-two W.2B/23 reverse flow engines and four W.2B/26 straight-through engines. The W.2B/23 became the RB.23 Welland engine, R for Rolls and B for Barnoldswick, and the W.2B/26 became the RB.26 Derwent engine. Rolls-Royce now called their jet engines after rivers, perhaps because of the flow of water, which was similar to the flow of air in the jet engine. There it was, the transition. At the height of the Second World War, the Rolls-Royce engine name change from birds of prey to rivers signified the next chapter in the history of Rolls-Royce, the jet age. By June 1943 the fourth prototype version, aircraft DG205/G, of the Gloster F.9/40 (G41) with 1,700 lbf (7.6kN) Whittle reverse-flow engines had flown successfully. Three other engines had flown in the Gloster F.9/40, the Rolls-Royce Derwent straight-flow engine, the Metropolitan-Vickers F.2 axial-flow engine and the de Havilland H.1 straight-flow engine. Like the numerical engine designations, the F.9/40 became the Gloster Meteor. In two years, the available engine thrust had doubled. The jet engine had started its own journey and service to mankind, both in peace and, unfortunately, in future 'Cold' and 'Hot' wars. In the 1950s the jet engine became universal in combat aircraft, closely followed in the 1960s by the civilian market.

RB.23 Welland
1,700 lbf (7.6kN)

The Welland engine had a centrifugal compressor, a double-sided impeller, and a single-stage axial-flow turbine. There were ten combustion chambers placed around the turbine section to produce a short but stubby engine. The reverse flow configuration had the combustion hot air flowing forward before reversing to exhaust through the turbine. The first examples by Rover had suffered from surging, uncontrollable engine rotation speed, and turbine failures. Adrian Lombard had moved with the engine from Rover and together with the expertise of Rolls-Royce's Stanley Hooker, the engine's problems were ironed out. By January 1944 two Wellands had been installed in the Meteor and by July the engines saw active service in a Meteor F.1 with No.616 Squadron RAF. The Welland was Britain's first production jet engine, 167 were built. The Welland would have a relatively short life as the Derwent overtook it in performance and, more importantly, reliability and replaced it in the Meteor.

It would be remiss not to mention some of the other more notable Rolls-Royce jet engines that continued the legacy of Rolls-Royce as an aero engine manufacturer into this century. Famous Rolls-Royce engine names such as the Derwent, Nene, Dart, Avon, Conway, RB.211, Olympus and Trent carried the company logo through good times and the bad times. The nadir was when Rolls-Royce was nationalised in 1971 by the British Government. It took sixteen years during the RB.211 and Olympus times for Rolls-Royce to become a private company again.

RB.26 Derwent
Derwent I 2,000lbf (8.9kN) Derwent V 4,000lbf (17.8kN)

The Derwent was Britain's second jet engine to go into production. Developed at the same time as the Welland, it was radically different in that it was a straight-through flow design, sequentially front to back, rather than the Welland reverse-flow. The air entered through a screened intake, was compressed by a centrifugal compressor, distributed by the diffuser to the can combustion chambers where fuel was ignited, and the resulting hot gases exhausted through the axial turbine. The design was done in secret as a parallel development to the Welland with Lombard guiding the work. Whittle resented this as he felt that all Rover's resources should have been assigned to his engine, the W.2B/23 (Welland).

The Derwent notably continued to support the Gloster Meteor until the 1960s and was installed in the experimental Avro 707, an aircraft designed to investigate tailless delta wing flight. Four Derwent engines were also installed in the Avro Canada C102 Jetliner prototype; it was just beaten by a few days by the de Havilland DH106 Comet to be the first commercial airliner to fly.

Rolls-Royce Nene
5,000lbf (22.24kN)

The Nene completed the trio of the Hooker/Lombard initial transition jet engines. A totally new design, it was the most powerful engine of that era. It saw limited applications in Britain, the Hawker Sea Hawk and Supermarine Attacker, and the Rolls-Royce Thrust Measuring Rig, the 'Flying Bedstead'. It was produced under licence in the USA as the Pratt & Whitney J42. Ironically, a reverse-engineered version, the Klimov VK-1, was installed in the famous Mikoyan-Gurevich MiG-15, a very successful Cold War fighter adversary. Now that would be an interesting story to read, how that came about! Apparently, the Labour Government gave its permission to sell Nene and Derwent engines to their 'ally', the Soviet Union; how things would change. By October 1944 the Nene produced 5,000lbf making it at that moment the most powerful jet engine in the world. The Nene was subsequently developed into the Rolls-Royce Tay, which had more success under licence in the US, the J48, and in Europe.

The Derwent V (above) was an 85 per cent scale Nene engine and, on 7 November 1945, established an official air speed record in a Gloster Meteor F.3 of 606mph (975kmh). The V had been specifically produced to fit in the Meteor. The Mk 10 Nene, as mentioned previously, was flown by the author in the Canadair T-33 Silver Star during Advanced Jet Training and as a No. 414 (Electronic Warfare) Squadron RCAF target and evaluation aircraft. The engine performed flawlessly for me during six years of Canadian winters and summer thunderstorms; thank you, Rolls-Royce!

RB.50 Trent
750 Shaft Horsepower (shp)

This engine should not be confused with the Trent series of high bypass turbofan engines. This engine was a derivative of the Derwent MkII and through a reduction gearbox drove a five-bladed Rotol propeller. In 1944 experimental work with the Welland engine had involved a spur-type reduction gear to measure shaft horsepower. It was the first turboprop engine in the world to fly. The first flight was the Derwent version of the RB.50 Trent in a Gloster Meteor in September 1945 to gain experience with jet and propeller combinations. The test flying programme was approximately 300 hours. The foundation was in place for further turboprop development for Rolls-Royce as it was well and truly established in the jet age with this initial experimental turboprop engine and the production centrifugal flow engines operating successfully. The next step in Rolls-Royce's aero engine history was the development of a viable commercial turboprop engine and the evolution of the centrifugal jet engine to the axial flow configuration.

RB.53 Dart

The RB.50 Trent led the way in the Rolls-Royce transformation out of the highly successful Merlin/Griffon era to the turboprop era. It was born in a time of pencils, drafting tables and slide rules belonging to designers sitting in large drawing offices. These same designers had to cope with rationing, wartime blackouts, travel restrictions and the ever present, although diminishing in the spring of 1945, threat of an air raid. It had been a long six years of wartime restrictions and it would take some time for life to return to normal. The designers did not have the luxury of Computer Aided Drafting or Design (CAD). The engine originated in Barnoldswick as the RB.53, which was to be the engine for a new RAF turboprop trainer. The importance of this engine is reflected in the fact that the work was moved to the main factory at Derby. It had a two-stage centrifugal compressor, seven combustion chambers and a two-stage turbine.

The Dart engine, RDa 1, first ran on a testbed in 1946 with the usual mixed results of any new design, primarily less power than calculated and expected. One of the driving forces behind the engine design was the Brabazon Select Committee's recommendation for a twenty-four-seat short-range pressurised turboprop airliner. The Air Ministry ordered a prototype of the Vickers V.630 Viscount and the government wanted to remain with the tried and proven centrifugal type jet engine. This perhaps was a bit short-sighted in hindsight but perhaps it was the best decision with the facts known at the time. The Dart first flew in a Lancaster in 1947, with the same lack of power results; British European Airways expressed a preference for the Bristol Centaurus piston-powered Airspeed Elizabethan. The future did not look good for the future of RR with their Dart engine.

In July 1948, the Viscount made its maiden flight, a thirty-two-seater now and plans for further seats. This was the first scheduled airline flight in the world by a turbine-powered aircraft. The two problems were addressed; weight had been saved by changes to the reduction gear and air intake casing and power had been increased by improving the efficiency of the compressor and turbine. The RDa.3 was officially rated at 1,400shp in March 1951 and was released as the production Dart Mk 505 in June 1952. The Viscount 630 had seen limited passenger flights in 1950 under a special certificate but it was the Viscount 700 series in 1953 that enjoyed the famous, for those times, quietness, comfort and the reliability of the Rolls-Royce Dart engines. The Viscount orders spread to North America and sixty airlines around the world. The Rolls-Royce engine and emblem became a common sight and soon other manufacturers sought the Dart to power their aircraft, the Fokker and Fairchild F27, Armstrong Argosy, Breguet Alize and the Handley Page Herald being examples.

Avon

The Avon was the first really successful Rolls-Royce jet engine and it could be equated to the Merlin on the piston side. It built on the success of the centrifugal engines as it pioneered the future of the Rolls-Royce axial flow engines. It was on the design table in 1945 as the Nene was proving its reliability and going in to operational service. Dr Griffiths had proposed the concept some years before and under Stanley Hooker, the research was moved from Barnoldswick to Derby. By March 1947 the engine ran, though not without major problems, which were subsequently solved. Failure to start, broken compressor blades and lack of acceleration all contributed to a less than desirable start of the Avon's life. Originally known as the AJ.65 – A for axial, J for jet and 65 for 6,500lbf – the engine initially adopted an RA nomenclature (R for Rolls-Royce and A for Axial?) to be followed by a numbered series in the hundreds. By August 1948, two-position guide vanes and the redesign of the compressor bleed ports cured some problems and allowed the RA.2 to fly installed in an Avro Lancastrian.

Development of the RA.2 continued through 1949. Early in the year, a pair of engines was sent to the English Electric Company for installation in the first Canberra. Later in the year, another batch was made and some of these were sent to Canada in January of 1950 for installation in the prototype Avro Canada CF-100 fighter. In 1971 the author flew the production CF-100 Canuck, which had the Canadian-built Orenda engines. On 28 August 1957 a specially prepared English Electric Canberra WK163 flew to altitude using its normal Rolls-Royce Avon engines and then used a rocket, the Napier Double Scorpion, to boost it to a record altitude of 70,308ft (21,430m).

The RA.3, the first production engine, had a two-stage turbine and was designated the Avon Mk 101 and assigned to the English Electric Canberra B.2 and B.6. This was the start of an illustrious career, retiring from the English Electric Canberra PR.9 in 2016, after sixty-eight years of Avon RAF service. The Avon powered a formidable collection of military aircraft, the Hawker Hunter, Supermarine Swift, Vickers Valiant, English Electric Lightning, de Havilland Sea Vixen and Fairey Delta 2. A four-Avon-powered de Havilland Comet C.2 flew the first scheduled transatlantic jet service in 1958. The Avon was also installed in the French Sud Aviation Caravelle. The Rolls-Royce Avons had followed in the historic footsteps of the Rolls-Royce Eagles by both engines being the first, albeit in different categories, to fly across the Atlantic Ocean.

The various Marks slowly increased in thrust until the Mk.300 series added reheat (afterburning). A RA.29 Mk301/2 (RB.146) produced 12,690 lbf (56,450 kN) normally and was rated at 17,110lbf (76.11kN) in reheat, a 35 per cent increase. Reheat allows for greatly increased power, thrust, only for short periods of time due to the high fuel consumption rate. It is mainly used in military aircraft for takeoff, catapult launches from an aircraft carrier and combat conditions. The Lockheed SR-71 Blackbird is an exception as it used the reheat to maintain its high supersonic cruise. Two civilian airliners also used reheat, the Aerospatiale/BAC Concorde and the Tupolev TU-144. Reheat was not new to Rolls-Royce. In 1944 a Rolls-Royce W.2/B23 engine with reheat was flight-tested in a Gloster Meteor. Reheat is simply injecting additional fuel into the tailpipe downstream of, after, the turbine. The reheat primarily increases thrust of the engine by accelerating the exhaust gases to a higher velocity. The reheat is controlled by the pilot and can be selected on and off as required.

The Avon engine achieved world air speed records in three different aircraft types between 1953 and 1956 and achieved a world land speed record in 1983 (*see* list, page 246). It is indicative of the Avon's success that more than 11,000 engines were built; a success story from start to finish for Rolls-Royce. Within five years of the Merlin ceasing production, Rolls-Royce had once again developed an outstanding engine to continue the legacy of producing superb aero engines. The transition from piston to axial flow jet engine was now completed, what was the next step?

Conway

The next step was the bypass, turbofan, engine and the first in the world to enter service was the RB.80 Conway. Yet again development had started in the 1940s, but the Conway had a relative short service life in the 1950s and 60s before the pace of development overtook it. It was the first turbofan and it was by Rolls-Royce. The name came from a Welsh river, the Conwy, adapted to the more acceptable Conway! The idea of bypass had been studied by Griffiths and Whittle but was put aside due to the demands of war. The success of the Avon allowed Rolls-Royce to return to the turbofan in the quest for more efficiency. The basic principle is that some air from the ducted compressor fan 'bypasses' the combustion chambers and turbine and joins the high velocity exhaust and contributes to the thrust. The ratio of jet thrust to fan thrust determines if it is a high-bypass or low-bypass type; if the jet thrust is higher than the bypass thrust, it is a low-bypass engine and vice versa. The Conway is a low-bypass engine with a ratio of 0.3; sixty years later, the Rolls-Royce Trent XWB is approaching 10.1. Most commercial engine applications are high-bypass for economy and quietness and most military are low-bypass, with or without reheat. The turbofan engine is considerably quieter than the pure axial flow type because of the slower moving fan air surrounding the high velocity exhaust air. Engine efficiency for modern-day air transport can be divided in two general groups, turboprops for low speed and ducted fans for most speeds of jet airliners. One of the earliest orders for the Conway engine came from Trans-Canada Airlines for the Douglas DC-8.

RB.211

The RB.211 is a family of high-bypass engines, originally run in 1969. The need for further engine efficiency, and airline profit margin, forced the development of high-bypass engines. Again Rolls-Royce was a leader, and the RB.211 was the result. The three-spool design elevated Rolls-Royce to global leader and the RB.211 family series was born. With its origins in the RB.178 and connected with the RB.203 and RB.207 for the Airbus 300, the RB.211 was scheduled to be available in 1970 but ran in to developmental problems. A disintegrating fan stage, higher costs and added weight all added to the engine's lack of progress. The projected costs had nearly doubled and now were greater than the selling price of each unit. Rolls-Royce went into receivership and the government nationalised it shortly thereafter. Rolls-Royce persuaded Stanley Hooker to come out of retirement and the engine was finally certified in April 1972. This did not help. Lockheed and the McDonnell Douglas DC-10 had a competitive advantage. The RB.211 was the prototype engine for the Lockheed L-1011 Tristar and was the only engine to power this aircraft. The tri-jet airliner was the third wide-body jet to enter commercial service after the Boeing 747 and the McDonnell Douglas DC-10. A total of 250 aircraft were built in two versions, standard length and a shortened version for longer range. Unfortunately, due to many circumstances, Lockheed was unable to make the Tristar profitable and withdrew from the civilian aircraft production business. The RB.211 turbofan engine could generate a range of power from 37,400 to 60,000 lbf (166 to 270 kN). An RB.211 version was installed in the Boeing 757 and helped Rolls-Royce to recoup some competitive production advantage. This was a time, as it still is, of intense competition for engine orders among the big three, Rolls-Royce, Pratt & Whitney and General Electric. The sales department even sold the engine to Russia to use in the Tupolev TU-204, the first western engines ever supplied to a Russian airliner. In 1990 the RB.211 proved itself an extremely reliable engine and was awarded 180-minute ETOPS (Extended Operations) on the Boeing 757, thereby opening the Atlantic for the Boeing 757. It would take twenty years for the RB.211 to evolve into the Trent series of engines.

Olympus

The Rolls-Royce Olympus, originally known as the Bristol B.E.10 Olympus, was the world's first designed/developed two-spool axial flow turbojet. It powered the Avro Vulcan in the 1950s and further design saw the engine used in the BAC TSR-2 supersonic programme; an Olympus 22R provided short periods of flight at Mach 2.2. A further, more widely known, claim to fame was that it powered the highly visible, in the public eye, BAC Sud Aviation Concorde SST. The reheated engine was called the Rolls-Royce/ Snecma Olympus 593. It was based on the Bristol Siddeley Olympus 22R with Rolls-Royce buying out Bristol Siddeley Engines Limited and joining with Snecma in further development and production. First test-flown in 1966, externally on an Avro Vulcan, the subsonic test produced 35,190 lbf (157 kN). Concorde 001 took off using reheat on 2 March 1969. The competitive Pratt & Whitney J58 powered the Lockheed SR-71 Blackbird at supersonic speeds and high altitudes. There were sixty-seven Olympus 593 built and sixteen preserved examples can be viewed, an impressive statement about how much the engine was admired.

Trent

The Rolls-Royce Trent series of engines is a direct development from the RB.211-524L continuing with the successful three-spool design. Thrust ratings have increased as per the demands of the aircraft manufacturer, in the range of 53,000 to 95,000lbf (240 to 420kN), to power various Airbus and Boeing models. The Trent has bridged the turn of the century, first run in 1990, and has made Rolls-Royce the second biggest supplier of civilian turbofan engines after General Electric. It was pleasing to see that Rolls-Royce had returned to naming their engines after rivers although I must admit that the RB.211 has a more powerful sound to it! The Trent River flows through the Midlands of England. When Rolls-Royce was privatised in 1987 it had to do something inexpensively to get back into the airliner market. The solution was to offer a series of engines developed and improved from the successful RB.211 for a range of aircraft. The three-spool arrangement offered a short, rigid engine with the ability to rotate the front fan at its optimum speed. This configuration allowed the engine to be individually scaled for individual performance and thrust requirements. Twenty-one years after its first flight, it has achieved the milestone of 100 million flying hours; some milestone, some engine. One of the important requirements of the Trent was to meet the ever-stricter noise regulations. The engine has plenty of components to adjust, change, redesign and improve the relationship between them as they strive to give the best competitive performance and carry Rolls-Royce forward to continue their success in the 21st century. Two men, Rolls and Royce, have left a legendary name known throughout the world. What a legacy in aero engine history from the Eagle to the Trent, so far!

Rolls-Royce (engine) world speed records

AIR

1929	R engine in S6	357.70mph
1931	R engine in S6B	379.05mph
1931	R engine in S6B	407.50mph
1945	Derwent V in Gloster Meteor	606.00mph
1946	Derwent V in Gloster Meteor	616.00mph
1953	Avon in Hawker Hunter	726.60mph
1953	Reheat Avon in Supermarine Swift	735.70mph
1956	Reheat Avon in Fairey FD2	1,132.00mph

LAND

1933	R engine Bluebird	272.10mph
1935	R engine Bluebird	276.82mph
1935	R engine Bluebird	301.10mph
1937	R engines Thunderbolt	311.42mph
1938	R engines Thunderbolt	345.49mph
1938	R engines Thunderbolt	357.50mph
1983	Avon in Thrust II	633.46mph
1997	Speys in Thrust SSC	763.03mph (first supersonic)

WATER

1930	R engines, Miss England II	98.76mph
1931	R engines, Miss England II	103.49mph
1931	R engines, Miss England II	110.22mph
1932	R engines, Miss England III	119.81mph
1937	R engine Bluebird	129.50mph
1938	R engine Bluebird	130.86mph
1939	R engine Bluebird II	141.74mph

Bibliography

Allward, Maurice, *Hurricane Special*, Shepperton: Ian Allan Ltd 1975

Armstrong, David, *How Not to Write a Novel*, Brixton: Allison & Busby Ltd 2003

Ashton, J. Norman, *Only Birds and Fools*, Shrewsbury: Airlife Publishing Ltd 2000

Barker, Ralph, *The Hurricats*, London: Pelham Books Ltd 1978

Bastow, Donald, *Henry Royce*, Derby: Rolls-Royce Heritage Trust 1989

Bingham, Victor, *Merlin Power*, Shrewsbury: Airlife Publishing Ltd 2003

Birch, David, *Rolls-Royce and the Mustang*, Derby: Rolls-Royce Heritage Trust 1987

Birrell, D. *FM159: The Lucky Lancaster*, Nanton: The Nanton Lancaster Society 2015

Birrell, D. *People and Planes: Stories from the Bomber Command Museum of Canada*, Nanton: The Nanton Lancaster Society 2011

Bowman, Martin W., *The Mosquito Story*, Stroud: The History Press 2011

Bracken, Robert, *Spitfire The Canadians*, Erin: The Boston Mills Press 1995

Bracken, Robert, *Spitfire II, The Canadians*, Erin: The Boston Mills Press 1995

Brady, John, *The Craft of Interviewing*, Cincinnati: Writer's Digest Books 1978

Bruce, G., *Charles Rolls Pioneer Aviator*, Derby: Rolls-Royce Heritage Trust 1990

Chisholm, Anne & Davie, Michael, *Beaverbrook A Life*, London: Pimlico 1993

Cotter, J. *Living Lancasters*, Stroud: Sutton Publishing Limited 2005

Cotter, Jarrod, *Battle of Britain Memorial Flight*, Barnsley: Pen & Sword Books Ltd 2007

Craighead, Ian, *Rolls-Royce Merlin*, Sparkford: Haynes Publishing 2015

Cull, Brian, *Hurricanes over Tobruk*, London: Grub Street 1999

Deighton, Len, *Fighter:The True Story of the Battle of Britain*, New York: Alfred A. Knopf, Inc. 1977

Dibbs, John and Holmes, Tony, *Spitfire Living Legend*, Botley: Osprey Publishing 1999

Edwards, Richard & Edwards, Peter, *Heroes and Landmarks of British Military Aviation: From Airships to the Jet Age*, Barnsley: Pen & Sword Books Limited 2012

Eriksson, Patrick G., *Alarmstart: The German Figher Pilot's Experience in the Second World War*, Stroud: Amberley Publishing 2017

Evans, Charles W., *Overhaul of Merlin Engines in India and the USSR*, Derby: The Rolls-Royce Heritage Trust 2009

Evans, Michael, *In the Beginning*, Derby: Rolls-Royce Heritage Trust 2004

Eyre, Donald, *50 Years with Rolls-Royce*, Derby: Rolls-Royce Heritage Trust 2005

Falconer, Jonathan, *Bomber Command Handbook 1939–1945*, Stroud: Sutton Publishing Limited 2003

Gallico, Paul, *The Hurricane Story*, London: Michael Joseph Ltd 1959

Garbett, M. & Goulding, B. *Lancaster at War 2*, New York: Charles Scribner's Sons 1980

Garbett, M. & Goulding, B. *Lancaster at War*, Shepperton: Ian Allan 1971

Garbett, M. & Goulding, B. *Lancaster*, Enderby: Promotional Reprint Company 1992

Goulding, B. & Taylor, R. J. A. *Story of a Lanc': NX 611* East Kirkby: Lincolnshire Aviation Heritage Centre 2010

Gray, Robert, *Rolls On the Rocks: The History of Rolls-Royce*, Salisbury: Compton Press Ltd, 1971.

Gunston, B., *Fedden*, Derby: Rolls-Royce Heritage Trust 1998

Gunston, Bill, *Rolls-Royce Aero Engines*, Wellingborough: Patrick Stephens 1989

Halliday, Hugh A., *Woody* Toronto: CANAV Books 1987

Harker, Ronald W., *The Engines Were Rolls-Royce*, New York: Macmillan Publishing Co., Inc. 1979

Harker, Ronald W., *Rolls-Royce From the Wings*, Oxford: Oxford Illustrated Press 1976

Harvey-Bailey, Alec, *Hives' Turbulent Barons*, Paulerspury: Sir Henry Royce Memorial Foundation 1992

Harvey-Bailey, Alec, *Rolls-Royce*, Derby: Rolls-Royce Heritage Trust 1983

Harvey-Bailey, Alec, Piggott, D., *The Merlin 100 Series*, Derby: Rolls-Royce Heritage Trust 1993

Harvey-Bailey, Alec, *The Merlin in Perspective*, Derby: Rolls-Royce Heritage Trust 1995

Harvey-Bailey, Alec, *The Sons of Martha*, Paulerspury: Sir Henry Royce Memorial Foundation 1989

Henshaw, Alex, *Sigh for a Merlin*, Reading: Cox & Wyman Ltd 1980

Hooker, S., Reed, H., Yarker, A., *The Performance of a Supercharged Engine*, Derby: Rolls-Royce Heritage Trust 1997

Hotson, Fred W., *De Havilland in Canada*, Toronto: CANAV Books 1999

Bibliography

Hotson, Fred W., The *De Havilland Canada Story*, Toronto: CANAV Books 1983

Iveson, Tony and Milton, Brian, *Lancaster The Biography*, London: Carlton Publishing Group 2009

Jackson, Robert, *The Hawker Hurricane*, Blandford Press: London 1987

Jacobs, Peter, *Hawker Hurricane*, Ramsbury: The Crowood Press Ltd 1998

Kirk, P., Felix, P., Bartnik, G., *The Bombing of Rolls-Royce at Derby*, Derby: Rolls-Royce Heritage Trust 2002

Lavigne, Michael & W/C Edwards, James F., *Hurricanes over the Sands Part One*, Victoriaville: Lavigne Aviation Publications 2003

Lewis, Damien, *The Dog Who Could Fly*, Atria Books: New York 2013

Lloyd, R. Ian, *Rolls-Royce, The Years Of Endeavour*, London: Macmillan Press, 1978.

Lord Montagu of Beaulieu, *The Early Days of Rolls-Royce*, Derby: Rolls-Royce Heritage Trust 1986

Ludvig, Paul A., *P-51 Mustang*, Hersham: Classic Publications 2003

Martin, Brian P., *Birds of Prey of the British Isles*, Devon: David and Charles 1992

Mason, Francis K., *The Hawker Hurricane*, London: Macdonald & Co. (Publishers) Ltd 1962

Mason, Francis K., *The Hawker Hurricane*, Bourne End: Aston Publications Limited 1987

McInstry, Leo, *The Hurricane*, London: John Murray Publishers Ltd 2010

McIntosh, Dave, *Terror in the Starboard Seat*, Don Mills: General Publishing Co. Ltd. 1980

Messenger, William E. and de Bruyn, Jan, *The Canadian Writer's Handbook*, Scarborough: Prentice-Hall Canada Inc. 1986

Mikesh, Robert C., *Excalibur III*, Washington D.C.: Smithsonian Institution Press 1978

Morgan, Bryan, *The Rolls-Royce Story*, London: Collins, 1971.

Moyes, Phillip, *Bomber Squadrons of the R.A.F. and their Aircraft*, London: Macdonald & Co. (Publishers) Ltd. 1971

Nahum, A., Foster-Pegg, R., Birch, D., *The Rolls Royce Crecy*, Derby: The Rolls-Royce Heritage Trust 2013

Nancarrow, F. G., *Glasgow's Fighter Squadron*, London: Collins 1942 Nockolds, Harold, *The Magic of a Name*, London: G.T.Foulis & Company, Ltd. 1961

Panton, Fred and Panton, Kate, *Man on a Mission*, Leeds: Propagator Press 2012

Price, Alfred, *Spitfire: A Complete Fighting History*, Enderby: Bookmart Limited 1991

Price, Alfred, *Spitfire MarkI/II Aces 1939–41*, London: Osprey Publishing 1996

Probert, Henry, *Bomber Harris His Life and Times*, London: Green Hill Books 2003

Pugh, Peter, *The Magic of a Name: The Rolls-Royce Story Part One*, Cambridge: Icon Books Ltd 2000

Pugh, Peter, *The Magic of a Name: The Rolls-Royce Story Part Two*, Cambridge: Icon Books Ltd 2001

Pugh, Peter, *The Magic of a Name: The Rolls-Royce Story Part Three*, Cambridge: Icon Books Ltd 2002

Quill, Jeffrey, *Spitfire A Test Pilot's Story*, Trowbridge: Redwood Books 1983

Rolls-Royce Group plc., *Rolls-Royce 1904–2004: A Century of Innovation*, London: Rolls-Royce Group plc 2004

Rubbra, A. A., *Rolls-Royce Piston Engines*, Derby: Rolls-Royce Heritage Trust 1990

Sherrard, Peter, *Rolls-Royce Hillington*, Derby: Rolls-Royce Heritage Trust 2011

Shores, Christopher, *The History of the Royal Canadian Air Force*, Toronto: Royce Publications 1984

Shores, Christopher and Cull, Brian with Maliziz, Nicola, *Malta: The Spitfire Year 1942*, London: Grub Street 1991

Spick, Mike, *Supermarine Spitfire*, London: Salamander Books Ltd 1990

Stokes, P. R., *From Gypsy to Gem*, Derby: Rolls-Royce Heritage Trust 1987

Sweetman, W., *Avro Lancaster*, New York: Zokeisha Publications 1982

Taulbut, Derek S., *Eagle*, Derby: Rolls-Royce Heritage Trust 2011

Williams, David E., *A View of Ansty*, Derby: Rolls-Royce Heritage Trust 1998

Wilson, G. A. A., *Lancaster Manual 1943*, Stroud:A mberley Publishing 2013

Wilson, G. A. A., The *Lancaster*, Stroud: Amberley Publishing 2015

Aircraft in Profile Volume 1, Windsor: Profile Publications Limited 1965

Aircraft in Profile Volume 2, Leatherhead: Profile Publications Limited 1965

Aircraft in Profile Volume 3, Windsor: Profile Publications Limited 1967

Aircraft in Profile Volume 5, Windsor: Profile Publications Limited 1967

Aircraft in Profile Volume 7, Windsor: Profile Publications Limited 1967

Aircraft in Profile Volume 9, Windsor: Profile Publications Limited 1971

Aircraft in Profile Volume 10, Windsor: Profile Publications Limited 1971

MAGAZINES

Aviation History, David Johnson November 1994

Aviation History, Radko Vasicek September 2002

CAHS Journal, The RCAF's UK-built Hurricane MkIs Carl Vincent Winter 2015

CAHS Journal, The RCAF's UK-built Hurricane MkIs Carl Vincent Spring 2016

CAHS Journal, The RCAF's UK-built Hurricane MkIs Carl Vincent Summer 2016

Canadian Aviator, Lisa Sharp, November/December 2014

Flight, H. F. King MBE 7 May 1954

Airforce Magazine, Cable. E Vol 40/No. 3

Index

Index

Merlin G 73, 191
PV-12 191
Mercedes DF80 54
Meteor engine 16, 189, 237
Meteor, Gloster 233, 238-40, 246
Metropolitan-Vickers 67, 143, 236, 238
Midland Hotel 31
Mitchell, Reginald J. 46, 91, 97, 105-9, 225
Moore-Brabazon, John 27, 34-5
Moor Lane, Royce memorial 44
Morrison, Greg 205-7
Mosquito, de Havilland 16, 90, 98, 129, 130-37, 173, 175, 181, 185-86, 192-4, 196-8, 201, 219-20, 226-7
Motor Car Club 23
Mustang 156, 158-64, 166-9, 173, 182, 184, 189, 195-96, 210-222, 226-7
Mustang Merlin engine prototype AL975-G 162
Mustang Mk X 16, 162
Mustang P-51 90, 154, 159, 163, 165, 169, 185, 219-21, 223
P-51B 164-6, 169, 185-6, 196
P-51D 166-7, 169, 186, 219, 222-3

N
Napier (D. Napier & Son Limited) 41, 67, 71, 73-4, 131, 178
Napier Double Scorpion 242
North American Aviation (NAA) 154-60, 162-4, 166, 169, 188
North American Mustang (*see* Mustang)
Northcliffe, Lord 28. 30, 33, 35-36, 42, 44

O
Olley, Maurice 41, 52, 54
O.R. 16 F.36/34 98

P
Packard
Packard Merlins 68, 75, 100, 104, 134, 135, 163, 166, 179-80, 182-90, 193-8, 207-8, 216, 218, 222, 227

Packard Motor Car Company 75, 173, 183
Panton, Andrew 202, 210, 212
Prototype K5054 109
Prototype K5083 98

Q
Quill, Jeffrey 108-11

R
RAAF 136, 175
Radar 119, 120-21, 125, 131, 135-6, 144, 146, 149, 174-5, 210, 225
RAF (Royal Air Force) 12-13, 47-8, 177, 48-9, 68, 73, 90, 96-7, 100, 102, 110-12, 114, 117-18, 121-2, 124-6, 13233, 135-7, 139-41, 150-53, 155-6, 159-60, 162-9, 174, 177-80, 202, 209, 219, 225, 229, 234-5, 237-8, 240, 242
RCAF 136, 155, 169, 175, 204, 213, 222, 240
Reduction gear 54, 58-9, 61, 71, 73, 78-80, 82-4, 229, 240
Reno Air Races 9, 11, 201, 216, 218, 229
Renault V-8 42, 52
Rolls, Charles 14-15, 20-22, 24-7, 29-38, 40-42, 94, 224
C.S.Rolls & Co. 29-30, 34
Killed in flying accident 38
Memorial at Shire Hall 38
ROTOL 68, 102, 115, 124, 182, 192, 240
Rouse, Martin 200, 219-20
Royal Automobile Club, RAC 30
Royal Aircraft Factory 67, 128
Royce, Henry 8, 14-15, 20-24, 26, 30-33, 35-45, 47-8, 50-55, 62, 66-9, 190, 224-5, 227
Born 44
Died 48
Royce Ltd 21, 23-4, 31, 34
RR engines
Avon 16, 18, 91, 239, 241-43, 246
Buzzard 16, 42, 44, 60, 71
Condor 42, 58-59, 70, 92
Conway R.B.80 239, 243